D1127548

Microscopes and their uses

Microscopes and their uses

Claude Marmasse

GORDON AND BREACH SCIENCE PUBLISHERS
New York   London   Paris

Gordon and Breach, Science Publishers, Inc.
One Park Avenue
New York, NY 10016

Gordon and Breach Science Publishers Ltd.
42 William IV Street
London WC2N 4DE, England

Gordon & Breach
7-9 rue Emile Dubois
75014 Paris, France

Library of Congress Cataloging in Publication Data

Marmasse, Claude, 1928-
    Microscopes and their uses.

    Includes bibliographical references and index.
    1. Microscope and microscopy. I. Title.
QH205.2.M37   535'.332   80-19617
ISBN 0-677-05510-2

For all those friends who
helped so much, and especially
for Jean Mozzanino and J. de la
Garanderie; and last but not
least, for Terry Freeman...

"First and foremost, the worker must know his tools. The microscope is not an automatic or self-sufficient instrument, but an accessory to the eye and brain of the investigator. As an instrument, it must be manipulated with skill; as an aid to observation, its limitations and optimum procedures for its use must be continually borne in mind...."

Emile Chamot & Clyde Mason.

FOREWORD

It is a great pleasure to acknowledge here the great
help derived from numerous comments and suggestions kindly
made by Dr. Walter C. McCrone. It does not follow that Dr.
McCrone endorses all the points of view developed by the author; or, for that matter, any of them. Microscopy is far
from fixed, and in a way the variety of viewpoints one en-
counters in the literature is a fairly good measure of the
vitality of this discipline.

PRINCIPAL NOTATIONS AND ABBREVIATIONS.

Arc   principal determination.

f   focal length.

F   first (front) focus or focal plane.

F'   secondary (back) focus or focal plane.

G   magnification.

IR   infrared.

n   refractive index.

NA, (NA) numerical aperture.

t   exposure time (in s).

UV   ultraviolet.

ε limit of resolution.

θ temperature (in °C).

λ wavelength.

'   element of the image space.

CONTENTS

neous use of various types of illumination can prove useful;
for instance the combination in brightfield microscopy of
transmitted and reflected lights often permits a dramatiza-
tion of depth effects. This technique is applicable not only
in chemical microscopy but also to many biological preparations.

*

*   *

There are tactics of microscopical research, that is
to say an art of turning the deficiencies or the weaknesses
of certain equipments to one's advantage. The first example
which will be mentioned here is due to Chamot and Mason who
suggested it in their description of a method for the determination of the refractive index of a liquid by a modification of the well-known duc de Chaulnes' procedure; in this
variant, one needs a mark (e.g., a scratch, a crossline) on
the base of the experimental cell and on which the microscope is focused. These authors suggested the use as reference, in place of the mark, of the image of a scale projected
by the condenser in the plane of the preparation. This is a
classical (although periodically forgotten) technique, but
what strikes the present author as particularly clever is the
advice given by Chamot and Mason to profit by the lack of

achromatism of the condenser (presumably they were using a condenser of the Abbe type, as it is customary in chemical microscopy) by focusing on an image of an intermediate tint, such as a pink one which is rather weak: under these circumstances the focusing becomes critically sensitive. In other words, the zero is established with a great accuracy, much greater than it would be attainable in the case of a scale projected with say a planachromatic condenser.

Another example concerns the halo so familiar to phase contrast microscopists; it is usually considered a nuisance. It is essentially due to a phase-change phenomenon and is therefore function of the difference between the refractive indices of the object observed and of its surrounding medium: the halo disappears when this difference is zero. Clearly this affords a method for measuring refractive indices, in particular of biological objects. In the hands of Barer, who used as immersion media isotonic solutions of bovine plasma albumin (Armour fraction V)[¶], it became a highly quantitative procedure, as is demonstrated by the data of Table 1.

¶   If only because of the residual salts they contain, proteins contribute to the tonicity of the solution. In the case of this particular albumin preparation, Barer established that 10% protein is equivalent to 0.1% sodium chloride.

TABLE 1.

Total solid concentrations expressed in % protein (g/100ml) of three varieties of *Corynbacterium diphteriae* (Klebs-Loefler *Bacillus*), after Barer. Note that on the basis of these data alone, obtained by phase-contrast microrefractometry, each variety is unequivocally distinguishable.

Variety	Solid medium		Liquid medium	
	range	mode	range	mode
mitis	31.5 – 39	33.1		33.6
intermedius	20.6 – 31.5	25	23 – 36.8	31.5
gravis	20.6 – 31.5	25		34.1

*

*   *

Microscopical technique is not a set technique and a
given instrument may be used in many different ways and for
various purposes. There is no reason to believe that the
flow of improvements and refinements which encompasses a few
centuries of microscopical research is going to dry up soon.
Many problems have not yet been solved and better ways of
handling microscopical observations are devised nearly ever-
yday. But one cannot depend on luck and an analysis in physical terms of the system under investigation will always
prove useful. A solid grasp of the physics of the microscope
apparatus and of the interplay of its components is required
before the investigator can let fly his imagination and at
one and the same time stand on solid ground when it comes to
interpreting his observations.

C H A P T E R   2

THE MICROSCOPE

2.1.   THE APPARATUS.

From a physical standpoint a microscope essentially is
- or functions as - a short optical bench. This is obvious
in the case of the older metallographic microscopes whose
components were mounted on an horizontal bench of traditional design, and is easily recognizable in the old British limb
and tripod design; the latter system which proved most versatile and provided a definite stability for a small weight was
finally superseded by the continental microscope with its
well known horseshoe base. The evolution of this pattern cul-
minated with the development of an instrument which can be
described as two aligned optical benches (one supporting the
bodytube and the other the condenser system) connected by a
C-shaped arm; this made possible the insertion of a larger
stage.

Mecanically, the horseshoe base microscope constitutes
a very satisfactory solution to the problem of supporting the

optical system of a microscope, as it provides a good stability at the expense of a reasonable weight, and has a large
clearance where it is most useful, that is to say just below
the substage condenser where one needs space to insert filters, etc..., and to adjust the orientation of the mirror.
Sturdy apparatus with a binocular attachment and a revolving
nosepiece with a capacity for four objectives have been constructed in this way but soon a very serious deficiency of
the mecanical design of the focusing system became apparent.
In what constitutes the most classical design, the focusing
function is divided between two independent mecanisms: the
gross adjustment is carried out by means of a rack and pinion
drive which controls the displacement of the bodytube, while
the fine adjustment depends upon a fine-thread screw (micrometric screw) controlling the position of a cam on which
bears a small buttress integral with the rack and pinion
drive unit. Under these circumstances, when the microscope
is in an upright position, the weight of the bodytube and of
all its attachments bears on the threads of the micrometric
screw. But the acceptable load on the fine threads of a precision screw is severely limited and this prohibited the use
of many desirable accessories. Subsequently designers adopted
the solution of moving the fine adjustment mecanisms on the

bodytube so that it only had to bear the weight of the nose-

piece (with its attachments); this is satisfactory from a

mecanical point of view, but the micrometric screw control

knob is now placed in a somewhat awkward position and this

design fell into discredit. Another solution, the one pre-

ferred today, was to be found in a system due to Anton van

Leeuwenhoek: the focusing stage.

A typical instrument of this class is built around a

short and rigid arm perpendicular to the base of the micro-

scope. A rack and pinion drive screwed to this arm controls

the vertical displacement of a bracket (the "arm") to which

is affixed the bodytube: this is the coarse adjustment[¶].

Note that the optical axis, which can be taken here as the

axis of the bodytube, should be parallel to the axis of

translation. On the other hand, a dovetailed bed is screwed

(independently of the preceding mecanism) on the fixed arm

to guide the displacement of a carriage bearing the stage:

this movement is governed (fine adjustment) by means of a

combination worm[§] and worm wheel. The rotation of the worm

is generally controlled by a knurled knob coaxial with the

¶   Provision is generally made for varying at will the
tension of this adjustment, so that accessories of quite
different weights may be used.
§   Already used in many horseshoe base microscopes.

control knob of the coarse adjustment.

But some difficulties arise when this design is used in conjunction with a trans-illumination procedure, as in the great majority of cases the distance between the apex of the main condenser and the plane of the preparation must be kept within rather close tolerance limits. In order to avoid the necessity of constantly refocusing the condenser, the latter is mounted on a rack and pinion drive integral with the focusing stage in such a way that the optical axis of the condenser is identical with the optical axis of the objective. It will easily be seen that this requirement, which is particularly stringent when the microscope is fitted with an aplanatic condenser, calls for parts machined with a high degree of precision.

In spite of its many advantages, the focusing stage is unlikely to eliminate completely the fixed stage as the latter is much more rugged. Note also that sophisticated micromanipulators (e.g., of the de Fonbrune pattern) are probably easier to use with the fixed stage apparatus for which they were specifically designed. On the other hand, it must be pointed out that through the combined use of a fixed (with respect to the base of the microscope) microtool and of a smoothly riding mecanical stage mounted on a focusing stage, one can perform a number of simple micromanipulations, such as the isolation of small elements in a not too crowded preparation.

Anyhow, as the number of basic components or of quasi-permanently fitted accessories steadily increases, the microscope becomes top-heavy (this is specially noticeable after the addition of a photographic camera) and soon the capabilities of a base of the horseshoe pattern are overtaxed. This led to redesign the base which nowadays is purposely made heavy (but hollow so that an illumination system may be inserted within) in order to provide a low center of gravity. The shape of the base is important too: a circular design has been used in some microscopes but it is rather inefficient and today research microscopes[¶] mostly incorporate a rectangular base plate, although T-shaped bases can also be encountered.

2.1.2.   Optical length.

In spite of a certain tendency in the contemporary technical and pedagogical literature to pass rapidly over the

---

[¶]   Traditionally the expression "research microscope" has been employed by manufacturers to denote an instrument particularly well designed and built. In a loose - but convenient - way, the expression implies the achievement of a high degree of versatility and stability; nowadays it seems to be used mainly in reference to heavy instruments.

concept of optical length, the latter parameter remains one
of the most basic characteristics of a microscope system,
and one with which, sooner or later, any investigator whose
activity leads him to do a little more than just to peer
briefly through a microscope will have to reckon.

In the case of a monocular instrument of the simplest
type (*i.e.*, constituted only by an objective and an eyepiece),
the optical length, also called projection distance[¶] or optical tubelength[§], is the distance between the back focal plane
of the objective and the front focal plane of the ocular. The
first plane is easily located as a stop[†] is most often placed
there; and for all practical purposes it can be taken as
passing through the middle of the knurled ring at the base
of the barrel; in the same way, a field stop is usually located in the front focal plane of an eyepiece.

It must be emphasized that only relatively small variations in the projection distance around its nominal value

---

[¶]   Note that this expression is much used also in microprojection and photomicrography to denote the distance between the projecting device and the screen or the photosensitive emulsion. The context generally makes it clear which
sense is intended.
[§]   It must be emphasized that this expression was coined
at a time when the main instrument of a microscopist was a
monocular microscope fitted with a drawtube. In that context, it proved valuable.
[†]   Or a phase-changing plate.

are permissible: this stems from the fact that in the object

space there exists but a narrow region (a thin slab extending

outside the front focal plane) within which the object under

observation must be located so that a reasonably sharp image

is obtained. The image of an object located outside this re-

gion either is virtual or presents so many aberrations that

it is quite unusable, but conversely, in critical microscopy,

aberrations can be minimized through a careful adjustment of

the projection distance.

The range of possible variations of the optical length

varies much with the type of optical components used, and

many objectives will be found for which variations of the

projection distance as large as ±10% do not appreciably de-

grade the quality of the image. This is a most fortunate cir-

cumstance for the user of a binocular head because in this

accessory a modification of the interpupillary distance is

correlated with a variation of the optical length. In a typ-

ical instrument the beam exiting from the objective is split

into two components of equal intensity, each of which is de-

flected towards a movable total reflection prism integral

with a tube in which an eyepiece is inserted: to adjust the

the interpupillary distance one varies the distance between

the two total reflection prisms. The design is generally

symmetrical, and therefore a variation of 2 mm of the interpupillary distance induces a variation of 1 mm of the optical
length. As the magnification of the objective is proportional
to the optical length, a change of the interpupillary distance will result in a noticeable change of the size of the
image[¶]. Furthermore the optical length along one of the two
paths (usually the left one) is independently adjustable so
that the instrument may be compensated for differences of
accommodation between the two eyes of the observer: this is
achieved by fitting one of the ocular-bearing tubes with an
helicoidal ramp which allows one to vary the relative position of the two eyepiece frontal planes.

When the microscope system includes optical elements in
addition to the objective and the eyepiece, the optical length
is calculated as $\sum_i n_i h_i$ (where h_i represents the thickness of
material of refractive index n_i); the summation is evaluated
along the path of the rays. This quantity rapidly increases
with n. It follows that the interposition of optical elements
of non negligible thickness will greatly increase the optical
length; in particular the insertion between the objective

¶   But the interposition of an intermediate stage deeply
affects the quantitative relationship between the interpupillary distance and the magnification; in modern instruments
the effect of the interpupillary adjustment is negligible in
all but extreme cases.

and the eyepiece of thick polars, such as Nicol prisms, is

likely to prove detrimental to the quality of the image un-

less the mecanical length is accordingly shortened.

From a practical standpoint, it is highly desirable

that the front focal plane of the various eyepieces used be

always located in the same position with respect to the top

of the tube in which a particular eyepiece is inserted: in

these conditions an interchange of eyepieces will not induce

any variation of optical length and the preparation will stay

in focus. Sets of eyepieces having this property are said to

be parfocal[¶]: one is dealing here with a problem of mecanical

construction and generally manufacturers provide series of

parfocal eyepieces. But some difficulties are likely to be

encountered if one has to use eyepieces of different commer-

cial brands: the only eyepiece modification which can be car-

ried out with any ease consists in raising the front focal

plane. This is done by slipping around the eyepiece barrel

¶   A quite different property is involved in the parfocalization of objectives: the distance of the plane of the
preparation (in focus) to say the nosepiece is constant for
a set of parfocal objectives. The parfocalization of unmatch-
ed objectives can be achieved by interposing between the
base of the objective barrels and the nosepiece receptacles
either washers (punched from a seet of tin-lead alloy or
machined from plastic or aluminium stock) or in extreme cases
small barrels fitted with RMS mounts: this results in a variation of the optical length.

a tightly fitting wide split-ring which is positioned on top
of the ocular tube.

2.1.3.   Centering.

As a microscope essentially is an optical bench, it is
necessary to align (or check the alignment of) the various
mecanical and optical components which constitute the system.
This operation, which is called centering or centration, is
easy to perform once the particular mecanical or optical component which defines the optical axis of the system has been
determined.

Let us note in the first place that the tolerance limits
with which the various centering operations have to be performed, widely vary according to the nature of the component
considered: in particular these limits are large in the case
of the eyepieces[¶], to such an extent that as a rule[§] no provision is made by the manufacturer for their alignment. In
the same way, except in the very important case of the petrographic microscopes, there is generally no possibility for

---

[¶]   This can be easily verified with a petrographic microscope. Focus on a standard preparation and by means of the
centering screws of the objective mount, displace the objective parallel to itself: the quality of the image will not
noticeably change, although the optical axes of the objective and the eyepiece do not coincide during all the course
of the manipulation.
[§]   With the exception of goniometer eyepieces which include their own centering device.

the user to modify the alignment of the objective in use.

Under these circumstances the optical axis of say a biolog-

ical microscope is defined by the axis of the objective and

for well constructed (and well cared for) instruments it can

be taken for all practical purposes as the axis of the body-

tube. It will be noticed that from a mecanical point of view

this throws a great burden on the revolving nosepiece which

should show only negligible play at the highest magnifica-

tions used and whose various threads should be rather tight;

this also explains why, when the alignment of the objective

is very critical, one resorts to individual dovetailed slide-

mounts.

Practically all research microscopes incorporate some

means to center the condenser: this is achieved for instance

by mounting it in a short sleeve which can be laterally dis-

placed with respect to the substage body by means of screws

working against a spring: two screws with knurled head in

the case of the condensers said to be centerable, and three

or four small set screws for factory-centered condensers.

The last expression should definitely not be interpreted as

meaning that no centering has to be done by the user. Screws

easily work loose, and as a matter of fact it will often be

found that the simple operation of removing the condenser

from its substage and of resetting it results in a slight

decentering. While the latter has generally no great impor-

tance in the case of an Abbe condenser used for brightfield

procedures, it may somewhat lower the quality attainable with

an aplanatic condenser in phase contrast microscopy.

The procedure used to center a condenser with respect

to the objective is based on the fact that the first focal

plane of the condenser and the secondary focal plane of the

objective are conjugate (see Fig. 6.2.2.); now the iris dia-

phragm of a brightfield condenser is usually located in the

first focal plane and the iris diaphragm of a phase contrast

condenser[¶] is located very near that plane. Under these con-

ditions, and assuming furthermore that the condenser is in

good mecanical conditions, the problem is reduced to posi-

tioning the condenser so that the image of the iris diaphragm

is centered in the back focal plane of the objective. This

image can be directly observed with the naked eye after re-

moval of the eyepiece. This method is especially convenient

in the case of monocular microscopes, but some difficulties

may be encountered with certain binocular attachments. In

¶   For centering purposes, a phase contrast condenser becomes a brightfield condenser by switching into the beam the
blank opening of the annulus-bearing turret.

that instance, the back focal plane of the objective can be observed with the help of a Bertrand ocular[¶]. Note that the same visualization procedure is used in brightfield micro-scopy to adjust to its proper value the diameter of the open-ing of the condenser iris diaphragm, that is to say to adjust the numerical aperture of the illuminating beam.

In instruments fitted with a non-centerable rotating stage (e.g., some petrographic microscopes), the axis of rotation of this component defines the optical axis of the system. Under these circumstances, the alignment of the microscope can be achieved in two consecutive steps: i) the objective is first aligned with respect to the axis of tota-tion of the stage, and then ii) the condenser is aligned with respect to the objective exactly as in the case of a fixed stage apparatus.

The centering of the objective requires the use of an eyepiece equipped with a crossline reticle, and is much fa-cilitated if a crossline preparation[§] is available[†]. Two

---

¶   See also 6.2.2.
§   Made for example by mounting in Canada balsam two human hairs: they should be set at approximately 90°. With high magnification objectives, thin hairs are preferable; they can be obtained from a young (true) blond female.
†   With some experience, one can dispense with a crosshair preparation. Use any well defined and easily recognizable point feature of a preparation: e.g., a particle, the apex of a crystal.

slightly different procedures may be employed depending
whether or not the rotating stage is equipped with a mecanical stage. In the first case, the "center" of the preparation
is brought to the center of the field, that is to say in coincidence with the center of the eyepiece reticle, by means
of the mecanical stage. Then the stage is rotated; when the
objective is not properly centered, the preparation center
will describe a circular arc. Estimate the position of the
center of the latter and through the use of the centering
screws of the objective mount, bring it approximately in coincidence with the center of the reticle. The manipulation
is repeated until the center of the preparation remains in
perfect coincidence with the center of the reticle during a
complete rotation of the stage.

When the preparation is held by means of stage clips[¶],
first of all its center is manually brought close to the
center of the field. If the instrument is perfectly centered
the preparation center will describe, during a complete rotation of the stage, a circle whose center coincides with
the center of the reticle. If this is not the case, one es-
timates the position of the trajectory center, brings it near

¶   Experienced microscopists will often perform this manipulation without bothering with stage clips.

the center of the field by means of the centering screws and repeats the process. No more than a few trials are required to center accurately an objective.

Centerable rotating stages are often used with objectives which are set in a fixed mount, e.g., a standard rotating nosepiece. In that case, the optical axis is defined by the objective in use and the alignment procedures are very similar to those used with non-centerable rotating stages: one merely substitutes a lateral displacement of the stage for a lateral displacement of the objective mount.

The centering of high magnification objectives may present some difficulties because of the narrowness of the field; in that case it often proves expedient to carry out a preliminary crude centering of the objective mount or of the stage with a low power (and wide field) objective.

2.1.4.   Standardization of microscope components.

The interchangeability of the optical components of a microscope is clearly a very desirable feature and the standardization - though limited - carried out by most manufacturers proves most convenient to the user.

Very early the standard number of 36 threads per inch was adopted for many optical assemblies; in particular it is still the standard thread number for objectives, and is call-

ed the RMS (Royal Microscopical Society) thread. Its profile
is most particular: it is not a Unified screw thread but has
a Whitworth profile with rounded crest and root. The main
characteristics of the RMS thread are: major diameter = 20.2
mm, pitch = 0.706 mm, flank angle = 55°. Note that even in a
metric world, the RMS thread is most likely to survive as it
is certainly less annoying to keep a special set of RMS
thread cutting tools than to condemn to obsolescence many
fine objectives.

What may be called a standard eyepiece[¶] is an eyepiece
with an outside diameter of *ca*. 23.2 mm (0.917"); there is
nowadays a strong tendency to manufacture eyepieces of this
size but components of a larger diameter may be encountered,
for example in a wide-field design. The adaptation of eyepieces of a non-standard diameter can be carried out by means
of brass or aluminium sleeves turned down from stock; this
may however entail a significant variation of the optical
length and therefore may induce some degradation of the quality of the image.

No standardization of condenser dimensions has yet been
achieved to any extent.

¶   In microscopy; astronomers for instance have another
standard: 31.75 mm (1¼") outside diameter.

2.2.   BINOCULAR OBSERVATION.

Due to the higher degree of sophistication of binocular microscopes, it is implictly assumed that they are better instruments and also (or consequently) that their systematic use entails less fatigue. This is a fallacy, and passing over the first point which shall be discussed later (see in particular 2.3.1.) we shall examine here some of the physiological aspects of the binocular mode of observation.

Let us note in the first place that during the course of monocular observation the body of the investigator is left a considerable latitude of movements; from a purely optical point of view, perfect conditions of observation will be met as long as the pupil of the eye remains in coincidence with the Ramsden circle[¶] (*i.e.*, with the exit pupil of the instrument). In fact small movements of the head and of the neck are likely to prove beneficial by eliminating an important source of crispation[§].

On the other hand, the position of the body with respect

[¶]   Also called Ramsden circle, eyepoint.
[§]   The discomfort experienced by some users of monocular microscopes is probably due, in many cases, to a faulty adjustment of the illumination system: for prolonged observations a softly lighted field is much to be preferred. Note also that, whether the mode of observation is monocular or binocular, a poor focusing of the eyelens on a crossline or a micrometer scale is likely to induce some fatigue.

to a binocular attachment, whose main value resides in its
capability of providing more comfortable working conditions,
is quite fixed. The benefits to be derived from the use of
this accessory will obviously be cancelled if the observer
sits in such a way that the neck muscles become cramped. In
particular the relative height of the microscope and of the
chair of the user is critical: a relaxed posture in which the
eyes naturally fall at the level of the eyepieces requires a
rather low stool or a high table. In the same way the fore-
arms should rest flat on the bench during prolonged observations; in fact this has become a "natural" position for the
microscopist as many instruments are nowadays designed so
that the main controls (*i.e.*, the knurled knobs controlling
the fine adjustment, and the longitudinal and lateral displacements of the mecanical stage) can be easily operated
when the operator properly faces the instrument.

Needless to recall that a binocular microscope is not
a stereoscopic instrument. As a psychophysiological counter-
measure to an uncomfortably flat image, various methods have
been proposed over the years for inducing a stereoscopic vision effect. They all rely on the introduction of some asymmetry in the intensity distribution of the two light beams
which independently reach a retina; for example:

1. if the distance between the axes of the eyepieces
    is less than the interpupillary distance, each (eye)
    pupil will function as an eccentric field stop for
    the corresponding light beam. Under these circum-
    stances the image received by each eye will exhibit
    a different repartition of light intensity;

2. the two beams of light which are sent to the eye-
    pieces can be differentially polarized by means of
    thin linear dichroic filters.

All these procedures are of a dubious value; as a matter
of fact the problem is not to modify artificially an image by
procedures which always entail a heavy penalty (a diminution
of effective aperture induces a loss of resolving power[¶], and
linear dichroic filters are definitely not achromatic) but to
obtain under well controlled conditions an image which originally contains the information sought. Rather than create a
false impression of relief, one should adopt conditions in
which the relief - if it exists - will be detected, e.g., by
selecting a specific illumination method.

¶   Either of the objective or of the eyes.

2.3.   SELECTION, SETTING AND CARE.

2.3.1.   Selection.

As a rule, it pays to buy microscope equipment from a reputable manufacturer, but without giving into the fetishism of brand names and keeping in mind that:

1. the size of a manufacturing company has little to do with the quality of its products. Some relatively modest outfits develop a habit of producing constantly improved materials;

2. most companies tend to excel in certain limited domains. It may be for example advantageous to buy objectives for polarization work from one company and phase contrast objectives from another one. There is therefore a lore of technical and scientific information which can be gathered only by experience and through personal contacts with seasoned microscopists.

Whatever may be the type of work planned, a well equiped microscopy laboratory should include at least one small monocular microscope fitted with an extendible drawtube and, if possible, with an inclinable arm. This type of apparatus has somewhat gone out of fashion, but is still available from a few manufacturers; and it is sometimes possible to locate a

so-called "student microscope" of old vintage and to put it
back into commission. Such an instrument will be found to be
a most valuable adjunct to a big research microscope as in
particular:

1. losses of light are minimum: this is a precious
    property when one has to examine weak images as it
    is the case in the study of the dichroism of bio-
    logical material and in fluorescence microscopy;

2. a drawtube is most useful in critical microscopy
    (e.g., for minimizing spherical aberration) or for
    the extemporaneous setting up of special optical
    configurations. In the latter instance, use can of-
    ten be made of the fact that the extremity of the
    drawtube is generally fitted with a RMS thread: one
    can for example screw a very low power objective to
    the drawtube: used by itself it affords limited ca-
    pabilities for macrophotography (see 5.2.5.); in
    conjunction with a standard optical configuration,
    it allows one to observe the back focal plane of
    the main objective (Bertrand system). In the same
    way, filters can easily be interposed across the
    beam, etc...;

3. a light-weight microscope is easily moved around

the laboratory and requires but little space. In
particular it is the ideal instrument to use for
following the time-honored recommendation of histologists and cytologists, *viz.*, to check frequently
under proper magnification the course of a staining
process;

4. an inclinable bodytube allows one to carry out ma-
    nipulations impossible to perform with a rigidly
    vertical instrument: such is the case in the com-
    parison of the refractive index of two liquids by
    the Schlieren technique. Furthermore an inclinable
    microscope is easily associated with many physical
    systems, and in particular can be employed as a
    versatile one-directional micromanipulator by affix-
    ing a microtool for example to the ocular tube.

2.3.2.   Location and setting.

Some thoughts should be given to the location of a big
research microscope, taking into account that it is neither
convenient nor advisable to move it frequently as it is a
fragile instrument whose weight may top 40 Kg. Although in
final analysis it is obviously the nature of the work to be
carried out which is considered when it comes to allocate
microscope space, a hard look should be taken to the general

environment; in particular:

1. dust, and specially abrasive dust, is deleterious.
   Also, chemical vapors, even in low concentrations,
   may be a problem. Note in this respect that the
   formaldehyde vapors which pervade the atmosphere of
   many a biological laboratory will quickly fog any
   photographic emulsion;

2. if it is contemplated to do much photographic work,
   vibrations should be avoided;

3. the possibility of rapidly darkening the room is a
   great convenience, as particularly for the efficient
   observation of faint images the eye must be dark-
   adapted. Two limits can be easily established for
   the amount of time required for this adaptation
   process: complete adaptation to darkness necessitates on the average 20 minutes if the eye was previously adapted to bright white light, and on the
   other hand the pupillary diameter does not reach
   its maximum value before 6 minutes have elapsed.
   However the time required to reach the state of complete adaptation to obscurity (dark vision) is reduced to a few minutes if, as it is well known to
   sailors and pilots, the eye was already adapted to

red light. It is therefore advisable that, near the
microscope at least, the light used to take notes,
etc..., be red and of a variable intensity. A small
flashlight with a red filter will also prove a useful accessory for checking various adjustments of
the microscope; it will often be necessary to paint
in white or yellow certain markings such as the
(commonly red) exposure time index on the knurled
ring of the photographic head shutter. Note also
that the dimming of the room lights will often result in a substantial decrease of stray light, which
is often useful for photomicrography and especially
advantageous for fluorescence microscopy.

2.3.3.   Mecanical care.

The paintwork of a microscope is best cleaned and protected by means of a thin layer of light wax (e.g., household
or car care preparation). Chrome and nickel plated parts are
cleaned with alcohol or a household polishing preparation
which does not contain abrasive particles, as the latter easily find their way in micrometric screws, rack and pinion
drives, etc... The same preparation can be used to clean brass
parts, but when the latter are very dirty a dilute solution
of ammonia in water is preferable: the parts are immersed in

a beaker with the cleaning solution until it turns blue. Then they are abundantly rinsed with water and polished; they may be protected with a light coat of varnish.

The advent of silicone and fluorocarbon compounds has considerably simplified[¶] the regular maintenance of the sliding and rotating parts of a microscope. Dovetailed beds are best treated with a fluorocarbon lubricant (applied with a small brush). Rack and pinion drives can be cared for in the same way, but if they exhibit some wear the use of a thick silicone grease (e.g., vacuum silicone grease) will often prove better. Some microscopes incorporate a ball-bearing in the focusing stage mecanism: it should be kept well packed with a standard lube or with silicone grease.

Iris diaphragms and related components present many problems of their own. As a general rule they are best run dry. In time the iris diaphragm of the main condenser will collect some dust: it can be removed with a camel hair brush. If this iris diaphragm does not operate smoothly, it can be lubricated with dry carbon, a very old recipe indeed. Gently

---

¶   In special cases when such coumpounds are not available (e.g., in field conditions), the dovetailed beds can often be run dry. As a general rule, one should be cautious in the use of penetrating oils, as they dry relatively quickly and form an abrasive paste with dust: then one has to keep oiling the mecanism over and over again.

run a soft lead pencil across the leaves and open and close
a few times the diaphragm to spread the lubricant. Then blow
away the excess of carbon.

The film cover and the shutter of a photomicroscope
(see Fig. 5.2.4.) deserve special consideration. Occasionally
the edges of the film cover may be wiped with a soft rag im-
pregnated with a fluid lubricant and then wiped nearly dry.
The photomicroscope shutter[¶] must be definitely kept dry;
even the use of carbon as a lubricant is prohibited because
some particles would be bound to fall on the deflecting mirror and finally on the intermediate stage (a component most
difficult to clean). In the same way one will be well advised
to keep away from the molybdenum greases advertised as lubri-
cants for the shutters of photographic cameras. A potential
problem with shutters is that of a mycelium growth on their
leaves; this phenomenon whose occurrence has been reported in
tropical and subtropical countries has for obvious consequence
to perturb in an erratic way the timing sequence of the device. No other remedy is known than to operate the latter
(dry runs) regularly when one has to work in a humid and hot
environment.

¶   The same comment applies to the case of the shutter of a
photographic camera fitted to a vertical microscope.

2.3.4.   Microscopes at sea.

Microscopy aboard a research vessel presents a few special problems; they are due not only to the movements of the ship but also to the pervasive corrosiveness of marine atmosphere.

The body of the investigator rolls with the ship, but with a noticeable phase lag; some people get used to it and learn how to anticipate slow rolls and to move with the microscope, though others seem never to be able to find their sea eyes. The best protection against accidents due to pitching resides in the use of eyepieces with rather high eyepoints. Many problems will be mitigated if the microscope is located along the lubberline, the microscopist facing the bow or the stern, and as close as possible to the ship center of gravity.

The microscope, as well as *all* its accessories, should be carefully secured to the bench, even when the sea is calm. The microscope can be fastened with C-clamps (which must be periodically checked for tightness); fixation by means of metallic bridles is probably easier to manage in the case of a microscope with a built-in illumination system. Large research microscopes generally have a base plate with three threaded holes (in view of securing the apparatus to the bottom of its

case when it is transported): this gives a convenient means
for bolting the instrument to the bench. Specially with binocular microscopes which are prone to accidents, care should
be taken that in case of a sudden jolt, the eyepieces do not
jump out of their tubes: a plastic bag secured by strong
rubber bands will prevent any mishap of this sort when the
instrument is not in use.

Corrosion is a very serious problem at sea, and one
which has not yet been satisfactorily solved as far as scientific instruments are concerned. The old stand-by's of sea-
manship: oil, paint and polish, still work best; delicate
unpainted parts (e.g., objective barrels, vernier scales)
can be protected with a thin coat of silicone grease, but it
must be recognized that this is at best a "messy" solution.
Common sense and reasonable care will help much to prevent
corrosion problems. For example, the instruments should be
wiped dry every day and before handling them, the investigator should rinse his hands in fresh water. In the same way
all the accessories which are likely to come in contact with
microscope parts, such as rags, tissue papers and lens paper,
should be kept in tight plastic bags.

C H A P T E R   3

OPTICAL COMPONENTS

3.1.   OPTICAL ABERRATIONS.

3.1.1.   Optical design.

In order to cope with the inherent limitations of the
optical systems he uses every day, not only must the microscopist be familiar with the optical characteristics of a
number of components, but he should be aware also of the type
of problems which face the optical designer and of the philo-
sophy with which they are handled. The state of the art has
been summarized by Herzberger:

> "The methods of correcting an optical
> system have not yet been developed in
> a satisfactory scientific manner.....
> ....
> A small change in a correction
> element usually has practically the
> same effects on all rays. Combinations
> of small changes can generally not

correct so-called "zonal" errors. *The*
art of the designer is to combine two
or more large changes in opposite di-
rections to obtain the desired result."¶

Such an approach can be discerned not only in the design
of individual components, for instance in the achromatization
of a doublet through the simultaneous use of crown and flint
glass elements, but also in the conception of the microscope
as a whole system; a familiar example of the latter instance
concerns the use of compensating eyepieces with semi-apochromatic objectives which are purposely left under-corrected
for chromatic difference of magnification while the former
are intentionally over-corrected: things can be managed so
that the overall chromatic difference is very small indeed.

3.1.2.   Geometrical aberrations.

The practical microscopist must be acquainted with some
results of the theory of aberrations and specially with its
terminology. Although the study of aberrations is a highly
specialized and difficult field of optics which cannot be
fully treated without introducing diffraction phenomena, the
existence of these aberrations can be easily shown in the

¶   Italics are Herzberger's.

following way. Snell's law for refraction can be written, with obvious notations, as:

$$n_1 \sin\theta_1 = n_2 \sin\theta_2 \qquad .$$

But such a formula is unmanageable when it is applied to the refraction of light rays through a lens, and its is customary to expand the sinuses by means of Taylor's formula. Thus one obtains as a first approximation:

$$n_1 \theta_1 = n_2 \theta_2 \qquad .$$

Gaussian optics is constructed with this approximation, but when the angles are not very small, one must take another term in the expansion of the sinuses, *viz.*, one uses the formula:

$$\sin\theta \simeq \theta - \theta^3/3! \qquad .$$

To this approximation corresponds an optics slightly different from Gaussian optics and which is called third-order approximation, Seidel approximation or Seidel optics. Roughly speaking, the difference between the Gauss and Seidel images of the same object can be measured by means of five independent parameters which are endowed with a geometrical meaning. These physical (geometrical) entities are called aberrations. They are:

1. spherical aberration which refers to the lack of
    union of rays which originate from a given point ob-

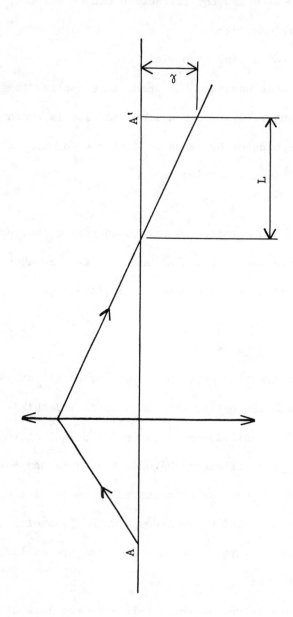

Fig. 3.1.2. Decomposition of spherical aberration into a longitudinal component (L) and a lateral component (ℓ); A' is the Gaussian image of A.

ject, at its Gaussian point image. This aberration
is customarily broken down into two components (see
Fig. 3.1.2.), a longitudinal (or axial) spherical
aberration and a lateral spherical aberration. When
spherical aberration is present, the image, observed
in a plane perpendicular to the optical axis of the
system, of a luminous point object is a circular
spot. Needless to say that this aberration plays
havoc with the power of resolution of an objective
or the usefulness of an eyepiece.

The total elimination of spherical aberration
in these components is extremely difficult, in particular because the optical designer is traditionally restricted to the use of spherical lenses which
are easier to grind in small sizes[¶]. On the other
hand it is technically feasible to parabolize small
mirrors[§] and this partially explains the everlasting
interest manifested in the development of sophisticated catadioptric systems;

2. coma which refers to the variation of magnification

[¶]   But aspherical lenses are at times used in condensers.
[§]   However the problem is most complicated: although a reflecting paraboloid surface presents the additional advantage
of being stigmatic, it is not coma-free.

of rays in pencils outside the paraxial[¶] region. In
other words, for off-axis points the magnification
is different for different parts of the lens. The
distribution of the light in the image of a point
is comet-shaped, which explains the name attributed
to that particular aberration. When the magnification is greater for the central rays than for the
outer rays, the coma is said to be negative; in the
reverse case, the coma is said to be positive.

This aberration can be fairly well eliminated;
in particular, one can design a coma-free spherical
lens and a doublet can be simultaneously corrected
for spherical aberration and coma;

3. astigmatism which refers to the difference between
the tangential and sagittal foci along a principal[§]
ray (*i.e.*, a ray which passes through the center of
the entrance pupil). The mean focus is the middle
point of these two foci, and the set of the mean
foci constitutes the mean focal surface. Note that,
except when the opposite is specifically stated,

¶   A paraxial ray is a ray originating on the optical axis
and making a very small angle with this axis.
§   Also called a chief ray.

the focal surface (or plane) is presumed to be the mean focal surface.

As a result of astigmatism, the rim and the spokes of a spoked wheel would be imaged on different surfaces (paraboloids of revolution); but note that there is no astigmatism for points lying on the optical axis. A system which presents this type of aberration is called astigmatic and a system which is free of astigmatism, either naturally or through corrections, is said to be stigmatic (it is an anastigmat).

The amount of astigmatism, measured by the distance between the tangential and the sagittal focal surfaces, is strongly dependent on the spacing of the optical elements as well as the location of the stops;

4. curvature of field which refers to the departure of the mean focal surface from a plane. Alternatively one speaks of the flatness of the field. This aberration is corrected by using the same means as in the case of astigmatism, that is to say by a judicious spacing of the optical elements and of the stops;

5.   distortion which refers to the displacement of an

image point from where it would be if the object

plane was mapped at constant magnification onto the

image plane. In the absence of distortion, the im-

age of a square is a square; if the image of a

square is a convex curvilinear quadrangle, the dis-

tortion is said to be of the barrel type while if

the image of a square is a concave curvilinear qua-

drangle, the distortion is said to be of the pin-

cushion type. This aberration can be minimized by

the adequate placement of a stop.

In many cases an optical system shows a combination of

these aberrations, but it must be emphasized that only sphe-

rical aberration may be associated with points lying on the

optical axis. A system corrected for both spherical aberra-

tion and coma is said to be aplanatic[¶].

Although Seidel aberrations are important, very impor-

tant, they are by no means unique and for the study of cri-

tical systems, e.g., objectives, one must take into consid-

eration aberrations of higher order than the third one. For

¶   This word is used at times in a (too) loose way to
qualify a corrected system in general. Note also that an
aplanat is a photographic lens (a symmetrical lens composed
of two achromatic cemented doublets).

instance it has become customary to include the results of
the fifth order approximation in the study of spherical aberration.

3.1.3.   Chromatic aberrations.

The Seidel aberrations briefly described above are defined for a quasi-monochromatic light. But as the refractive
index of a glass is wavelength dependent, many optical properties and parameters are wavelength dependent too (*i.e.*,
are color dependent). A simple example will show the order
of magnitude of the phenomenon: let us consider a thin lens
of 16.0 mm focal length for blue light (around 486 nm). If
it is made of ordinary crown glass, its focal length for red
light (around 656 nm) will be approximately 16.3 mm; this is
to say that the red and blue foci are widely apart. This
type of chromatic aberration is called longitudinal (or axial) chromatic aberration; there exists too a lateral chromatic aberration [see Fig. 3.1.3.(a)].

The correction of the chromatic aberrations of an optical system is called its achromatization. The meaning of
this term must be carefully weighted as it does not necessarily refer to a complete color correction, which may even
be theoretically impossible in certain cases. For example it
can be demonstrated that two separated thin lenses achroma-

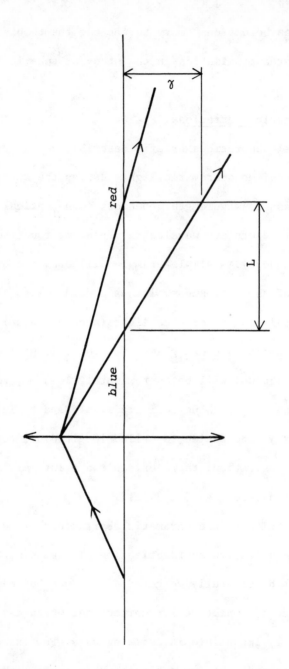

Fig. 3.1.3.(a). Decomposition of a chromatic aberration into a longitudinal component (L) and a lateral component (ℓ).

tized with respect to the power[¶] of the system are also achromatized with respect to the size of the image, but are not achromatized with respect to the position of this image. It should be noted that this achromatism condition is simply that the separation distance be equal to half the sum of the focal lengths: this design is widely used in eyepieces, notably of the Huyghens and Ramsden types.

In practice, the achromatization of the focal length of a system does not necessarily guarantee the elimination of longitudinal chromatic aberration because the position of the principal planes [see Fig. 3.1.3.(b)] is itself color dependent in general. In order to obtain a substantial reduction of the chromatic aberrations, one has to achromatize a number of the optical system parameters. This usually accomplished by associating lenses which have different dispersions, that is to say whose refractive index varies with the wavelength in different ways, A classical combination involves crown and flint glasses (see Table 3.1.3.); the residual chromatic aberration is often called the secondary spectrum.

¶ *I.e.*, the inverse of the focal length of the equivalent lens. When the focal length is measured in meters, the power is expressed in diopters.

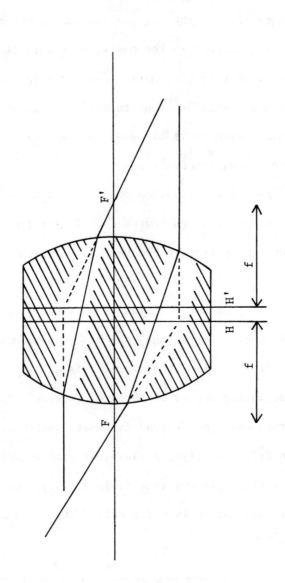

Fig. 3.1.3.(b). Some cardinal elements of a thick lens in air. F: front (primary, first) focus; F': back (secondary, second) focus; H: primary principal plane; H': secondary principal plane; f: focal length. The intersection of a principal plane with the optical axis is known as a principal point. The knowledge of the principal planes (planes of unit magnification) and of the foci allows one to construct the gaussian image of any object. For an example see Fig. 3.1.3.(c).

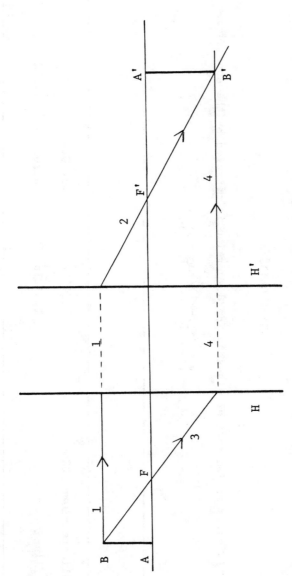

Fig. 3.1.3.(c). Construction of the Gaussian image A'B' of the object AB [see Fig. 3.1.3.(b)] when the principal planes (H and H') and the foci (F and F') are known. The rays used in the construction have been sequentially numbered to facilitate the interpretation of the figure.

TABLE 3.1.3.

OPTICAL GLASSES.

Characteristics of two optical glasses. The dispersion is measured by the Abbe number ν :

$$\nu = (n_D - 1)/(n_F - n_C).$$

Glass	n	n	n	ν
Borosilicate crown (BSC-2)	1.51462	1.51700	1.52264	64.5
Dense flint	1.61216	1.61700	1.69201	7.7

An optical system is said to be achromatic (and is called an achromat) when it is corrected for two colors[¶]; similarly an optical system is said to be apochromatic (it is an apochromat) when it is color corrected for three colors[§].

3.1.4.   Testing for aberrations.

The accurate measurement of the parameters of an optical system is a difficult operation which necessitates highly specialized and costly equipment[†], but it is relatively easy for the microscopist to evaluate qualitatively the performances of the system he uses. He should be able to recognize the occurrence of various types of aberration and to estimate their magnitude and he should know also how to look for them. And this for a very practical reason: he essentially makes use as parts of an optical system[×] of individual com-

---

[¶]   But note that in a too loose sense it may mean any achromatized system. The context sometimes indicates if the author refers to the strict or to the broad acception of the term.
[§]   The reader may well wonder what happens with systems corrected for a greater number of colors. Systems achromatized for four colors are called superachromats; it has been demonstrated that superachromatic lenses are achromatized for all intermediate wavelengths.
[†]   E.g., the Twyman and Green modification of the Michelson interferometer.
[×]   There are possibly some exceptions as in the projection of a real image of an object by the objective alone in photomicrography, or in conoscopic observations by the Lassaulx method; but it could be argued that these..... (contd. p.52)

ponents (condensers, objectives, etc...) which, however sophisticated they may be, still have in most cases a number
of small aberrations. When such elements are combined in a
system, their aberrations may be magnified or tend to compensate each other; the preparation itself can introduce
some aberrations. Under these circumstances it can be but
the responsability of the sole microscopist to see to it
that the aberrations of the total system are kept to a minimum, and also to estimate the importance of the residual aberrations.

A small set of test preparations is advantageously kept
at hand; they will be found invaluable not only for evalu-
ating the performance of the optical system but also for adjusting various illumination systems (e.g., darkfield condensers) and as test slides for a variety of photomicrographical procedures.

The standard test preparations of older microscopists
have fallen into disuse, perhaps too hastily as they are convenient to check rapidly the resolving power of an objective
(and also the quality of an illumination system). They were

... (contd. from p. 51)   exceptions are more apparent
than real as one generally has to reckon with the presence
of a condenser, of polars, ...

principally preparations of diatoms mounted in air; it is of
interest to note that later studies by electron microscopy
have confirmed the extreme constancy of certain of the dimensions of diatoms. *Pleurosigma angulatum* shows two sets of
parallel striae distant by 0.5 μm; these two families of
striae make an angle of 58°. *Surirella gemma* shows striae
distant by 0.44 μm; at high magnification ovoid pearls (major diameter: 0.44 μm; minor diameter: 0.38 μm) should be
visible.

The major disadvantage of preparations of diatoms mounted in air is that they diffract an appreciable fraction of
the incident light. At times this is an objectionable characteristic, e.g., when it comes to test phase contrast illumination systems. For this purpose the diatoms are best
mounted in a medium with a high refractive index. Note in
this respect that a preparation of *Globigena* ooze is very
convenient: elements of widely different sized are present,
and therefore many different tests may be carried out with a
unique slide. The sample is well dried and washed several
times with an organic solvent. A fraction is spread on a
slide in a large drop of xylol. Before the latter has completely evaporated, the larger elements are rearranged under
low magnification with the help of a dissecting needle and

the preparation is mounted in Canada balsam. Alternatively
the elements may be individually picked out and transferred
to a slide smeared with viscous Canada balsam.

Some well stained histological sections should also be
available; thin sections (approximately 4 μm thickness) cut
from paraffin blocks are convenient, though ultrathin sections of thickness ranging from 0.09 to 0.12 μm (their interference color is pale gold) cut from plastic blocks might
prove to be better yet. There is certainly nothing critical
about the type of tissue and the staining method to be used:
sections of rat liver and kidney stained following a classical hematoxylin-eosin procedure have proved perfect for the
purpose considered, in brightfield microscopy, but any thin
preparation which includes finely delineated details will
serve equally well.

Many small objects easily available to the microscopist
are suitable for the preparation of test slides, in particu-
ar pollen and starch grains (to be mounted in a liquid medium) and pigments. A permanent preparation of the latter may
be made as follows: a commercial white paint containing ti-
tanium dioxide is diluted with an appropriate organic solvent; evaporate a couple of drops on a slide and wash gently
with xylol. Then mount in Canada balsam.

It must be emphasized that the scrutiny of an image for the detection of possible aberrations should not be limited to those cases when the preparation is illuminated by means of a brightfield system: for example spherical and chromatic aberrations are often prominently displayed in darkfield illumination.

When spherical aberration is present (and it could well be due to a coverslip of incorrect thickness), no focusing manipulation will enable one to obtain a sharp image of the preparation. In brightfield microscopy and specially when small grains are examined, the optical effect known as the Becke line phenomenon may be strong and should not be confused with a manifestation of spherical aberration. The Becke line is a narrow bright halo which borders the image; it is more clearly noticeable in an object bound by approximately vertical planes and axially illuminated. The main property of the Becke line is that it moves towards the medium of higher refractive index when the microscope is focused upwards, and conversely that it moves towards the medium of lower refractive index when the microscope is focused downwards[¶].

[¶]   This phenomenon if of an exquisite sensitivity and consequently is much used in optical crystallography for the determination of the refractive indices of ...   (contd. p56)

Stigmatism is best checked with a preparation of small platy objects: an elongated image (observable by varying the focusing) is indicative of astigmatism. Very small objects endowed with a cylindrical or spherical symmetry are ideal for detecting coma: asymmetrical images reveal the presence of coma aberration. Distortion and flatness of field can be simultaneously studied by using as preparation a thin square or rectangular network: a special one may be constructed for this purpose¶ or a good quality hemocytometer can be used with low and medium power objectives. Most high dry power objectives cannot be tested with such a counting cell because of their too small working distance but no problems will be encountered with immersion objectives. A preparation of well formed platy or needle-like crystals can be used for detecting a strong distortion aberration.

The presence of chromatic aberrations is usually detected by the apparition of a blurred image following a change in the color of the illuminating light. However the colored halos seen around small particles surrounded by a liquid medium may be due to an entirely different phenom-

---

... (contd. from p.55)   a crystal by an immersion procedure: differences of refractive indices of the order of 0.001 to 0.0001 can be recognized.
¶   See 7.2.3.

enon: generally the optical dispersion of a liquid is greater than that of a solid. Under these circumstances, if the
refractive indices of the solid and of the liquid are matched for the yellow-green portion of the spectrum, the solid
will have a higher refractive index for red light and a
lower index for blue light. Then the image of a particle
will show an inner reddish halo and an outer bluish one:
the observation of these halos is often facilitated by a
slight refocusing of the microscope.

Note that all these aberrations may be appreciable in
the outer parts of the field only; under these circumstances
the central part of the image will be quite usable.

3.2.   OBJECTIVES.

3.2.1.   Functions.

Most modern objectives are calculated to form a real
image of the object at an approximate distance of 160 mm
from their back focal plane, but there are exceptions of importance. In the first place, certain European manufacturers
have produced microscope objectives, intended for biological
research, calculated for an optical length of 180 mm. Also,
many metallographic objectives have been calculated for
projection distances of, e.g., 180 mm, 200 mm and infinity.
In the latter instance, they are sometimes called tele-

centric[¶]: such objectives are useful when the light beam
must pass through thick optical elements, such as a Glan-
Thompson or a Nicol prism, which induce a substantial variation of optical length. However the importance of telecentric
objectives has been lessened by the introduction of the intermediate stage and by the generalized use of thin linear
dichroic filters as polarizing elements.

The image formation process may be quite complex and in
particular may involve the recombination of light beams which
have experienced different histories, e.g., in variable phase
contrast methods using a polanret system and in interference
microscopy; this obviously throws a heavy burden on the
objective. At times also, it is not the image of the preparation which is observed: for example in conoscopic procedures one examines interference figures in the back focal
plane of the objective.

Objectives are often used as condensers when a small and
intense light beam is required, as it is the case with the
Zsigmondy ultramicroscope. With prism vertical illuminators
and with reflecting illuminators of the Beck type, the objective simultaneously plays a double role: in the formation

---

¶   *Stricto sensu* a system is said to be telecentric on the
image side when its exit pupil is at infinity.

of the image and as a condenser. Note however that in certain situations the objective performs alternatively these
two functions: for example the same quartz objective may be
used firstly to focus an UV beam on a biological preparation,
in order to induce a well localized microlesion, and after-
wards to observe, in brightfield with a trans-illuminating
visible light, the consequences of the biological aggression.

3.2.2.   Classification of objectives.

Microscope objectives can be classified in a number of
ways, mainly according to:

1. their focal length, or equivalently their magnifi-
    cation;

2. their numerical aperture (NA). This is an essential
    characteristic as the limit of resolution of an ob-
    jective is inversely proportional to its NA;

3. the type of achromatization carried out. Customarily
    one distinguishes uncorrected objectives (a rarity
    today), achromats, semi-apochromats and apochromats;

4. particular corrections of geometrical aberrations.
    In especial, an aplanatic objective achromatized
    for two wavelengths has been called a planachromat;
    the use of this term calls for some caution as at
    times it is applied to condensers to denote an

achromat corrected for flatness of field (or a wide-
field achromat);

5. the nature of the optical material used. In particular one speaks of quartz objectives, which are
used for work in the UV region, and of fluorite objectives. The latter most often contain components
of both regular optical glasses and fluorite, and
generally are semi-apochromats;

6. for immersion objectives, the nature of the immersion fluid;

7. special mecanical features of the objective. For
instance all a class of objectives, usually called
metallographic objectives or short-mount objectives,
have a purposely shortened barrel so that the interposition of a vertical reflecting illuminator, with
which they are currently used, does not appreciably
modify the optical length of the system. Other objectives have an iris diaphragm which enables the
investigator to modify at will their numerical aperture. A fairly recent innovation is the development
of spring-mounted objectives: the various optical
groups are mounted in an independent cylinder which
is positioned relatively to a standard objective

barrel by an helicoidal spring, The latter is norm-
ally extended; in case of an accidental contact
between the front lens and the preparation, for
instance because of a mistake in focusing, the independent cylinder recedes within the barrel, and the
extent of the damages is somewhat limited. This is
a very convenient feature but it would be unwise to
rely systematically on its protective virtues rather
than to focus slowly and carefully;

8. special optical features. There are numerous objec-
    tives of this kind. In the first place, there are
    all the phase contrast objectives, which in their
    great majority include a phase plate located in the
    back focal plane. Two other classes of objectives
    at least must be mentioned: strain-free objectives
    whose use is quasi-obligatory for critical work in
    polarization microscopy and objective for fluores-
    cence studies. The latter objectives are mainly
    characterized by the use of a special cement to glue
    the lenses of the various optical groups: this sub-
    stitution is made necessary by the fact that Canada
    balsam, the conventional optical cement, is strongly
    fluorescent;

TABLE 3.2.3.

Some of the most frequently used equivalences between nominal focal length and magnification (for an optical length of the order of 160 mm). The accuracy of these equivalences can be ascertained by taking as an example the case of the 2 mm objective: in its classical version it has a focal length of 1.91 mm and $1/12'' \simeq 2.1$ mm.

Nominal focal length in mm	in "	Approximate magnification
32	4/3	5
16	2/3	10
8	1/3	20
4	1/6	40
3	1/8	60
2	1/12	100
1.6	1/16	125

9. their intended use. As it has been already mention-
    ed, short-barrel objectives are often referred to as
    metallographic objectives. In the same vein, certain
    40X objectives are often referred to as hematologi-
    cal objectives (hema or haema, in short): they have
    a relatively large working distance, which allows
    one to use them with the standard hematologocal
    counting chambers, and some of them are corrected
    for the thick coverslips (0.4 mm) which must be used
    with these cells as the regular coverslips of thick-
    ness 0.17 - 0.18 mm are too flexible.

3.2.3.   Focal length and magnification.

It results from an elementary theorem of Gaussian optics
that the focal length of an objective and its magnification
for a given projection distance (measured from the back focal
plane) are related as:

magnification = projection distance ÷ focal length .

A list of equivalences between the focal length expressed in
millimeters and inches, and the magnification, will be found
in Table 3.2.3.

Although the magnification of an objective does not say
much about its capabilities, this parameter is often used
when one wishes to describe rapidly an optical configuration.

This is justified by the fact that in practice certain characteristics of an objective of given focal length do not vary too widely around typical values: for example a 10X objective may be expected to have a numerical aperture between 0.25 and 0.45, with a working distance between 10 and 6 mm, while the numerical aperture of a 40X objective is likely to fall between 0.60 and 0.75, with a working distance in the range 0.6 - 0.2 mm.

3.2.4.   Resolving power and numerical aperture.

The existence of a relationship between the resolving power[¶], or limit of resolution, of an objective and its numerical aperture [defined as (NA) = n sin u/2, where n is refractive index of the object space and u the acceptance angle of the objective] was brillantly demonstrated by Ernst Abbe; this physicist derived a numerical expression relating these two quantities under certain hypotheses. Later Abbe's work was extended and the influence of a number of parameters was systematically studied. It should not come as a great surprise that slightly different expressions are obtained if different initial conditions are assumed; this explains in part the fact that apparently contradictory formu-

---

¶   Note that a more "powerful" instrument has a smaller resolving power.

lae are listed in physics and microscopy texts; on the other hand it must be realized that some serious errors have crept (and have been perpetuated) in a number of textbooks. For example, it has been suggested many times to write the expression of the resolving power of an objective as

$$1.22 \; / [(NA)_{obj} + (NA)_{cond}]$$

in order to emphasize its dependence upon the numerical aperture of the condenser. This is certainly a most respectable wish. However one cannot break arbitrarily a formula into various components just on the basis of feelings and the preceding formula is physically unsound. As a matter of fact it is the ratio of the numerical apertures of the condenser and of the objective which appears to be important[¶] and not the sum of these two quantities.

A critical analysis of the concept of resolving power of an objective, that is to say of the smallest distance which may separate two points so that they can be perceived as distinct, reveals that it is much more complex than has

---

[¶]   See 4.4.5. In microscopy literature the expression

$$[(NA)_{obj} + (NA)_{cond}]$$

is known as the "working NA". The persistence of this formulation is all the more surprising when one considers that *traditionally* microscopists have described the settings used in term of the ratio of the numerical apertures!

been thought for a long time and that some uncertainties are
hidden under the apparent simplicity of the definition. This
stems from the fact that the results of the calculation of
the resolving power of an objective depend on:

1. the shape of the aperture;

2. certain physical characteristics of the object, in
    particular its periodicities;

3. certain characteristics of the light, mainly its
    wavelength, its degree of coherence and the symmetry
    of the illumination (obliquity). Note in particular
    that the light which exits from a condenser is par-
    tially coherent;

4. the mode of observation. In the case of visual ob-
    servations one generally utilized a criterion pro-
    posed in 1879 by Lord Rayleigh for the evaluation of
    the performances of prism and grating spectrometers:
    two spectral lines are considered just resolved when
    the central maximum of one of them coincides with
    the first minimum of the other[¶].This criterion which
    is convenient and adequate as far as visual observa-

[¶]   When the two spectral lines are of equal intensity, the
ratio of the intensity of the maxima to that of the observed
minima is often referred to as the peak-to saddle ratio.

tions are concerned is somewhat arbitrary and proves
to be too restrictive when photometric methods are
employed. This point has been studied in great details in the case of the Michelson stellar interferometer: a higher sensitivity is obtained if one
analizes statistically the output of two photode-
tectors (intensity interferometry) than if one works
visually.

A photometric system at the fingertips of the
microscopist is a photographic emulsion: it is well
known for example that the limit of resolution of
an optical interferometer is often decreased when
one resorts to photographic detection methods (obviously the subsequent processing of the emulsion
- in particular its development - is critical). It
follows that the power of resolution attainable in
photomicrography is intrinsically higher than that
attainable by visual observation.

The scope of the problem is best ascertained by comparing the results obtained in three classical situations.
Case 1. When the light emerging from a self-luminous object
is incoherent, the limit of resolution ε according to the
Rayleigh criterion is given, in the case of a circular aper-

TABLE 3.2.4.

Airy units (A.u.) in μm as a function of the numerical aperture; these values are calculated for the mercury green line.

NA	A.u.	NA	A.u.
0.10	3.3	0.85	0.39
0.15	2.2	0.90	0.37
0.20	1.7	0.95	0.35
0.25	1.3	1.00	0.33
0.30	1.1	1.05	0.32
0.35	0.95	1.10	0.30
0.40	0.83	1.15	0.29
0.45	0.74	1.20	0.28
0.50	0.67	1.25	0.27
0.55	0.61	1.30	0.26
0.60	0.55	1.35	0.25
0.65	0.51	1.40	0.24
0.70	0.48	1.45	0.23
0.75	0.44	1.50	0.22
0.80	0.42	1.55	0.21
		1.60	0.21

ture, by the relation

$$\varepsilon = 0.61\lambda/(NA)$$

By convention the latter quantity represents one Airy unit
(see Table 3.2.4.).

Case 2. When the light emerging from a self-luminous object
is totally coherent, it has been theoretically shown and experimentally verified[¶] by Abbe that, in the case of a grating
as object, and with a circular aperture, all the diffraction
spectra contribute to the formation of the image. In practice, spectra of high or very high orders are always excluded
and do not participate in the formation of the image since
the objective always has a too limited aperture. A calculation presented by Born and Wolf leads for the limit of resolution to the expression

$$\varepsilon = 0.77\lambda/(NA) \qquad ,$$

that is to say to $\varepsilon = 1.26$ Airy units.

In this computation, one considers the image of two
distinct points: the distribution of light in that image is
due to the superposition of two Airy diffraction patterns
centered on the Gaussian image of the two object points. The

¶   In fact the experimental verification was not carried
out with a self-luminous object: the illumination was provided by a condenser of low numerical aperture.

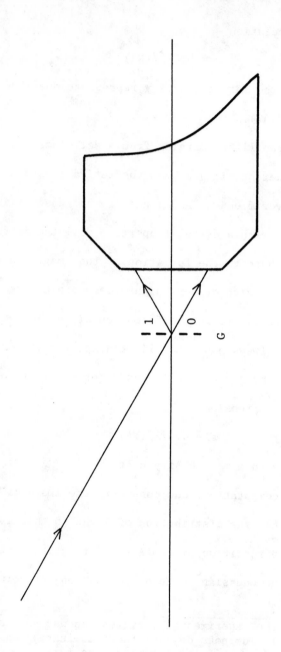

Fig. 3.2.4. Oblique illumination. In addition to the transmitted beam, the objective collects one diffracted beam of order 1. G: grating.

two images are considered resolved when the relative intensity of light at their midpoint (with respect to the intensity at one of the points) is the same as it was in case 1.

Case 3. With a grating illuminated by coherent light, a limiting condition for the formation of an image is that, in
addition to the diffracted beam of order zero (transmitted
light), one diffracted beam of order one enters the objective; at the extreme, this condition is fulfilled when the
incident beam is oblique relatively to the microscope axis
(see Fig. 3.2.4.). The smallest grating period for which
this is realized is a measure of the limit of resolution of
the objective; it comes:

$$\varepsilon = \lambda/[2\times(NA)] \qquad ,$$

that is to say $\varepsilon = 0.82$ Airy unit.

Let us first consider the result of case 3, as it is
the smallest estimate of the limit of resolution. This is
the classical expression of the resolving power of an objective such as it appears in most textbooks of physics. Its
value for the practical microscopist is somewhat ambiguous
because:

1. very rarely indeed has the investigator to examine
    periodic structures which act as diffraction grat-
    ings;

2. but this formula reveals the definite increase in
    resolving power which is afforded by the use of ob-
    lique illumination (which can be easily verified by
    examining diatoms for example). However oblique il-
    lumination procedures should be used with caution
    as often spurious images are generated: for instan-
    ce it can be shown theoretically and demonstrated
    experimentally that, if the objective is allowed to
    collect only diffracted beams of the second order,
    the periodicity of the grating is apparently dou-
    bled. Therefore it is in principle advisable not to
    use oblique illumination alone but to observe the
    changes in the image when one passes from geometric-
    ally axial to oblique illumination: in this way pos-
    sible artifacts will be easily detected.

The results of cases 1 and 2 are not substantially different and, taking into account the fact that the light
which exits from a condenser is only partially coherent, it
appears reasonable to adopt as estimate of the attainable
limit of resolution of an objective the value of one Airy
unit, as far as visual observations are concerned. However
the effective limit of resolution may prove much greater as
in addition to the influence of the shape of the object

(the estimation adopted is really based on the consideration of a two-pinholes structure), the adjustment of the numerical aperture of the condenser can be a decisive factor[¶].

A quantity which is related in an essential way to the resolving power of an objective is the useful magnification of the microscope. In the case of visual observations this parameter is the ratio of the resolving power of the eye to that of the objective. The first term depends on a number of physiological - and eventually pathological - variables: in particular the sensitivity of the eye to the light used, the age of the investigator, whether he is rested or fatigued, etc... It follows that the useful magnification of a microscope is not a fixed quantity (except for an ideal observer). But one can derive an estimate accurate enough for practical purposes. The classical rule, said of the 1000(NA), which states that the maximum useful magnification is equal to a thousand times the numerical aperture of the objective, is not useless as a guide-line but it provides a rather crude estimate and what is worse, the arguments presented for its justification are quite unconvincing. The latter instance is obviously to be expected from any calculation which involves

---

¶   See 4.4.5.

the eye but is short on physiology. As the argument touches upon some basic points which have a direct bearing on the use of a microscope, we shall dwell to some extent on the subject.

In the classical theory which leads to the 1000(NA) rule, the attention is focused on the separation distance of two luminous points; therefore one does not consider the rods (which are mainly involved in color vision). In these conditions, an estimation of the limit of resolution of the eye is made as the distance between two adjacent cones in the *fovea centralis*. This procedure is highly questionable because, although this distance likely plays an important role, there is no evidence that it is the unique parameter of the problem. As a matter of fact, it is very doubtful that this is the case:

1. there is no reason to believe that, even if two electrophysiological signals were simultaneously recorded in two adjacent cones of the *fovea centralis*, they would be recognized and interpreted by the cognitive structures of the brain as proceeding from two different external sources[¶]. All the expe-

---

[¶]   From this point of view, it is interedting to note that in the cat at least, the existence of lateral connections between some of the bipolar cells of the retina has been reported.

rimental evidence accumulated so far is strongly
suggestive of a stochastic nature for many visual
processes;

2. this model implicitly assumes a signal-to-noise ratio much too high to be credible;

3. in this model, either one attributes to the cones
an importance unjustified, or one forgets an essential parameter: this stems from the fact that although the *fovea centralis* is the most sensitive
part of the retina, it is also the one with the
lowest density of cones (*ca.* 4×10^3 per mm^2);

4. even if it were correct that the excitation of two
adjacent cones regularly generates signals interpreted as initiating in two different retinal zones,
it still would be a poor model of what happens when
a microscopist observes a preparation because (except perhaps in the ultramicroscopical study of
some colloids), he is automatically looking for a
structure in the image. Again this basically involves a statistical mecanism, as do all pattern recognition processes; this is certainly not contained
in an unsophisticated model which takes into account
only two adjacent cones.

Under these circumstances, although its simplicity is
undeniably attractive, it seems preferable to forget - at
least for the time being - this kind of approach and to use
an integrated index of performance, that is to say a parameter which is not solely based on one type of data, but takes
into account the whole chain of phenomena involved in vision
and in particular includes the brain recognition processes.
For the purpose at hand such a measure of visual acuity is
conveniently offered by the smallest variation of length
which can be perceived. A trained observer easily reads without the help of a vernier a millimetric scale to 1/10 mm; after a day's work, when he is tired, his visual acuity will
decrease a little but he will certainly remain able to read
the scale with a precision superior to 2/10 mm. Therefore as
an overall measure of the capability of discrimination of
the eye of the microscopist, or in other words as an estimate
of the limit of resolution of his eyes, we shall adopt the
value 3/20 mm. Taking the limit of resolution of the objective as one Airy unit, one obtains for the useful magnification of the microscope in green light:

$$G_u = 15 \times 10^{-5} \times (NA)/(0.61 \times 546 \times 10^{-9})$$
$$= 450 \times (NA) \qquad .$$

The variations of the useful magnification with the

wavelength of the light used are fairly important: assuming the value 3/20 mm for the limit of resolution of the eye o-ver the whole visible spectrum, one finds that $G_u/(NA)$ var-ies from 615 (extreme violet) to 352 (extreme red). But on the other hand visual acuity is color dependent in a way which strongly varies from observer to observer so that any general statement about regions other than the green-yellow part of the spectrum (which is the zone of maximum sensitiv-ity of the human eye in bright light) is bound to be either imprecise or incorrect.

Magnifications greater than the useful magnification are said to be empty. As a rule. it is advisable to work with some degree of empty magnification (provided that one is con-scious of the fact) because this makes for more comfortable conditions of observation. An extreme example is furnished by a procedure used for measuring the melting point of aniso-tropic microcrystals: under crossed polars an isolated crys-tal is brought to the center of the field and oriented in position of maximum brightness, with help of a compensator if the birefringence is small. The hot stage is slowly heat-ed and the image observed: it disappears when the crystal melts. In such a procedure, the indication sought, the end point, is the transition bright → dark (in the absence of a

compensator) or a change of color (when a compensator has
been inserted in the optical train). The pattern of the image
is of no interest here[¶] and as one generally wishes to use an
objective of large working distance (and therefore unfortunately of low NA), most of the magnification is produced by
the eyepiece. For example, one could use an objective [3X,
0.10 NA] with a 25X eyepiece; with an optical length of 160
mm, the total magnification will be 75 while the useful magnification estimated for green light is of the order of 45
only. But one can go one step further as we care little if
aberrations (even spherical aberration) are present and use
a projection distance of the order of 20 cm: the total magnification now becomes close to 94 and may prove sufficient
to observe a very small crystal. The image will certainly not
be comparable to what could have been obtained with a 10X objective followed by an 8 or 10X eyepiece, but the working
distance will be greater by a factor 3 or 4, a very desirable
feature when a hot stage is run at high temperature.

3.2.5.   Markings on the objective barrel.

Nowadays most manufacturers engrave on the barrel of

[¶]   But this is not necessarily the case. With the configuration described one could, for example, study allotropic
transformations.

their objectives *some* characteristics. However one must de-
plore a serious lack of standardization in this domain.

Some objectives are characterized only by a number, for
instance "No. 5" or "8". More frequently modern objectives
bear engraved their nominal magnification for an optical
length of 160 mm; this is indicated either by a number stand-
ing alone (usually very conspicuously) or by a number followed by "X".

The numerical aperture is indicated either by a number
alone[¶], as for instance in

Wild Fluotar

0,45

or an abbreviation such as "NA" or "N.A." will appear close
to the appropriate number.

Some objectives bear the indication of the optical
length (in mm) for which they have been calculated[§], and/or
the thickness (in mm) of the coverslip for which they have

¶   According to the geographical location of the manufacturer or that of the intended market, a decimal number is
written with a point or with a comma.
§   When it is not indicated on a modern objective, it may
be presumed to be equal to 160 mm.

been corrected. These two data may appear as a sequence of
two numbers separated by a slash. For example, there is a
Zeiss objective which bears the markings:

$$3$$

160/-

and which is a 3X objective, calculated for an optical length
of 160 mm and which can be used without any coverslip. In
some cases the thickness datun is preceded by a letter. For
instance there is a Wild objective which bears the indications:

Wild Fluotar

$$40$$

0,75

d = 0.17

and which is a 40X objective with a numerical aperture of
0.75, calculated for an optical length of 160 mm and a coverslip thickness of 0.17 mm. When the coverslip thickness is
not indicated, it may be assumed to be 0.18 mm unless the objective falls in one of the following categories:

    1. its nominal magnification is equal to or less than

10;

2. it is a short-mount objective which is to be used
    without a coverslip;

3. it is an oil-immersion objective (such a component
    generally bears the indication "Oil imm." or "IMM-
    HOM");

4. it is an hematological objective which should be
    used with a coverslip of 0.4 mm thickness.

Sometimes the type of correction of the objective is indicated or suggested; an example of the second instance is
offered by the trade name "Fluotar". This refers to a fluorite (and therefore presumably - but see 3.2.6. - semi-apochromatic) objective made by Wild in Switzerland. At times,
the aspect of an objective can suggest its type of correction
as modern achromats are, as a rule, much slimmer than their
semi-apochromatic and apochromatic counterparts.

The markings of phase contrast objectives are highly
variable, as it could be expected in a domain whose terminology is not yet completely fixed. However all these objectives
bear a notation of the type "Ph"; the manufacturer'literature
must generally be consulted to decipher the meaning of the
annexed signs and letters. This is not a serious problem as
it is relatively easy to determine to which broad class a

given phase contrast objective belongs. In the classification

prevalent today and which will be used here[¶], a phase A ob-

jective is such that the conjugate area (that is to say the

area optically conjugate of the condenser annulus) is absorb-

ing. Therefore the corresponding phase plate, which is usual-

ly located in the back focal plane of the objective, will

show a dark ring; it will be a light ring in the case of a

type B phase contrast objective. To determine if the objec-

tive functions with positive or negative contrast, recourse

must be had to the use of standard preparations. Note howev-

er that the barrel of phase contrast objectives of type A-

is at times painted in black.

Near their tip, certain objectives of modern fabrication

have a ring whose color is indicative of the nominal magnifi-

cation (for instance, light blue for 10X objective). This

feature is helpful in a teaching laboratory.

3.2.6.   Achromatization of the objectives.

The classification of achromatized objectives follows

the standard classification[§] but an extra-class, that of the

semi-apochromatic objectives, is introduced.

The achromatic objective is the war-horse of microscopy.

[¶]   See 6.2.1.
[§]   See 3.1.3.

In the absence of other specific information, it may be presumed to have been corrected for the C and the F lines. As a rule, this type of achromat shows better geometrical corrections in the green region of the spectrum and therefore, for critical microscopy, should be used with the corresponding filter. However other types of achromatization than the CF one are possible; in particular many quartz objectives designed for use in the UV region (and not to be confused with the quartz objectives used for fluorescence work) are achromatized for a specific UV line and a visible one: this feature is very convenient as it allows one to focus visually the microscope and then to record photographically the UV image or to exploit it by microphotometry.

Although apochromats by different makers vary in the degree of correction brought to their various geometrical aberrations, they may be expected to be fairly aplanatic. One of the advantages of the apochromats is that their actinic and their visual foci coincide: this enables one to focus with great accuracy for photomicrographical work. But because of the large number of lenses necessary for correction purposes (for a 2 mm objective, ten lenses in five groups are not uncommon), apochromats absorb much light, more than the corresponding semi-apochromats and achromats.

The class of the semi-apochromats is defined in a negative way: a semi-apochromat is not an apochromat, that is to
say is not chromatically corrected for three colors, but it
is much better corrected than a regular achromat. They are
often called fluorite objectives because of the relatively
high percentage of fluorite elements used in their construction, but this is hardly a specific characteristic as fluorite lenses are also used in true apochromats. Objectives
which include fluorite elements are not, as a rule, well
suited for polarization work but there are exceptions and
each objective should be individually tested; also, as fluorite is often fluorescent (the fluorescence color is variable,
depending on the nature of the impurities of the material)
such objectives are generally unsuitable for UV and fluorescence work, except obviously if the latter is carried out
with a darkfield trans-illumination procedure.

3.2.7.   Immersion of objectives.

The immersion of an objective is said to be homogeneous
when the refractive index of the immersion fluid is equal to
that of the front lens (see Table 3.2.7.). In a number of
cases the immersion is purposely inhomogeneous: for instance
with the water-immersion objectives used in zoology, certain
glycerine-immersion objectives specially designed for use

with darkfield illumination condensers, not to mention the immersion ultramicroscope! Note also that in optical crystallography one occasionally has to use for the determination of the refractive indices of very small crystals immersion fluids for which the system is inhomogeneous: in that case the image may be seriously degraded and yet prove usable.

Homogeneous immersion has long been preferred for critical microscopy because the front lens of the objective can easily be made to perform in nearly aplanatic conditions. In fact, since more than a century[¶], the first two lenses of a homogeneous immersion objective function in strictly aplanatic conditions for a small region centered on a point located on the axis, as it is shown on Fig. 3.2.7. It must be emphasized that any definite departure of the index of refraction of the immersion fluid from its nominal value will result in refraction phenomena at the interface immersion fluid - front surface: then the geometrical conditions which define the aplanatism conditions will not be satisfied anymore and a noticeable degradation of the quality of the image is to be expected.

The quality of an immersion fluid varies from batch to

¶   This system was described by Amici in 1844.

TABLE 3.2.7.

IMMERSION FLUIDS.

Refractive index at room temperature of the most frequently used immersion fluids.

Fluid	n_D	Immersion homogeneous with:
Water	1.333	
Glycerine	1.455	fused quartz
Cedarwood oil	1.515	light crown glass
α - monobromonaphtalene	1.658	flint glass

batch. This is probably at the origin of the controversy which developed some years ago about the use of sandalwood oil in fluorescence microscopy; in the same way, glycerine, even highly purified by standard methods, can contain traces of fluorescent contaminants. Therefore each batch of immersion fluid intended for use in fluorescence microscopy should be carefully checked by the investigator. Note also that cedarwood oil which was once the immersion fluid most used in brightfield, darkfield, polarization and phase contrast microscopy, is not completely stable. As a matter of fact it has been claimed to improve optically with age. Today much more stable synthetic oils are available from the microscope manufacturers.

Too little attention has been paid to other physical properties of an immersion fluid, although its viscosity and its surface tension are of great practical importance to the microscopist who routinely uses immersion procedures. An oil too fluid and too mobile "runs away" and air bubbles can easily become entrapped in a too viscous oil. In a general way, a lighter oil is more convenient for immersing the objective and a more viscous one for immersing the condenser. As both viscosity and surface tension decrease when the temperature increases, the use of a viscous (at room temperature) is advisable with warm preparations.

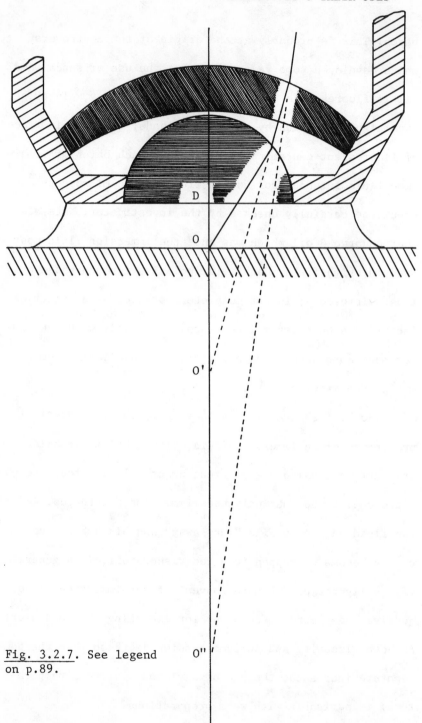

Fig. 3.2.7. See legend
on p.89.

Fig. 3.2.7. (contd. from p.88). Schematic cross-section of
a homogeneous immersion objective (Amici's design). The
first lens is hemispherical (center D, radius r and refractive index n): it gives a virtual aplanatic image O' (DO' =
rn) of the object point O located on the axis at the distance r/n from the center. This lens is followed by a convergent meniscus whose first surface is centered at O' and
whose second (rear) surface is such that O' is one of its
aplanatic points; its second aplanatic point, O", is therefore an aplanatic image of O.

3.2.8.   Working distance.

The working distance of an objective, sometimes called
its frontal distance, is the distance between the apex of
the front surface of the objective and the preparation. In
many circumstances it is desirable to have objectives (and
condensers) with a large working distance, for example for
microdissections and micromanipulations, for chemical manipulations and when hot or cold stages are employed, for per-
fusion studies of cells and tissue cultures, etc...

However there exist severe limitations on the range of
possible values for the working distance of an objective of
given magnification. This stems from the fact that the object

under observation must be located near (outside) the front
focal plane so that a real image is obtained at a convenient
distance from the back focal plane; but the focal length is
measured from the primary principal plane [see Fig. 3.1.3.
(b)], and the latter passes within the objective: thus the
front focal plane is close to the front surface of the objective. The optical designer can move the primary principal
plane forward, but at the expense of other desirable characteristics. More specifically, the working distance decreases
when the numerical aperture increases; also, an apochromat
will generally have a smaller working distance than the corresponding achromat. Finally large working distances are syn-
onymous with loss of aplanatism and of flatness of field. As
a consequence the working distance of a refracting objective
is always much smaller than its focal length.

Reflecting objectives on the other hand have a relatively large working distance: of the order of their focal length.
Objectives of the Burch type which make use of reflections on
two mirrors are unfortunately very bulky and probably for
this reason have found little favor. Systems in which reflection and refraction phenomena are associated are somewhat
slimmer. The two inherent qualities of catadioptric systems,
viz., their achromatism and their large working distance,

have attracted the attention of many opticians and it is
likely that, once their bulkiness is overcome, they will
prove most convenient for a number of investigations.

3.2.9.   Coverslip thickness and spherical aberration.

The problem of the minimization of spherical aberration
is extremely complicated: the optical system to be corrected
is composed of the objective itself and also of all what is
present between the front lens of the objective and the object examined, that is a layer of some fluid (air or immersion fluid) and often a coverslip. Obviously most of the
calculations of the optical designer refer to the image space
as only there can an estimation of the magnitude of the aberration be made; however a discussion of what happens in the
object space is much easier and proves adequate for the purpose at hand.

Let us consider in the first place the effect of the insertion of a coverslip on the image furnished by a short-
mount objective (that is to say an objective intended for use
without a coverslip) in the case of a point object trans-
illuminated with a quasi-monochromatic light [see Fig. 3.2.9.
(a), diagram on the left]: all the rays coming from the object O and collected by the objective form an image assumed
to be a point. We now interpose a coverslip (diagram on the

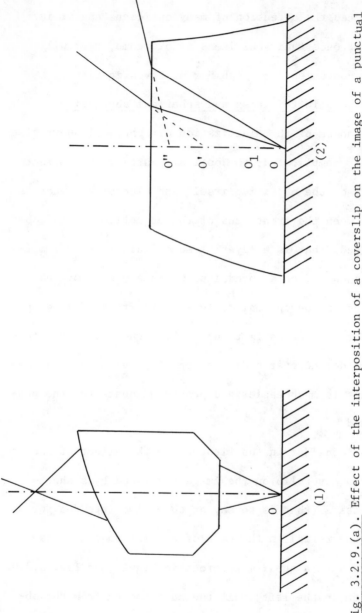

Fig. 3.2.9.(a). Effect of the interposition of a coverslip on the image of a punctual object O given by a short-mount objective. In (1) the image is punctual; in (2) the rays seem to come from O_1, O', O'', etc....

right) of thickness h and refractive index n. Along the axis, the image of O through the coverslip is O_1. Two non-paraxial rays have been traced on the diagram: they are refracted at the glass-air interface and seem to come from O' and O" respectively. It follows that, for the objective, the object is no longer a point but has become a short segment. In general, the image of this segment will not be a point; this is to say that the interposition of the coverslip has induced some spherical aberration.

The spread (ε) of the points O', O", etc... is clearly a measure of the phenomenon. The maximum displacement is obtained for the cone of the rays which can be but barely collected by the objective. Using the notations of Fig. 3.2.9. (b) [p.94], one obtains by writing the sine law in the triangle O'DD':

$$(h - OO')/\cos\theta = DD'/\sin\theta \qquad .$$

We also have
$$DD' = h \tan\alpha \qquad .$$

Taking into account the two relations
$$\theta = Arc \sin(NA)$$

and
$$n \sin\alpha = \sin\theta \qquad ,$$

one obtains:
$$OO' = h\{1 - [1 - (NA)^2]^{\frac{1}{2}}[n^2 - (NA)^2]^{-\frac{1}{2}}\} \qquad .$$

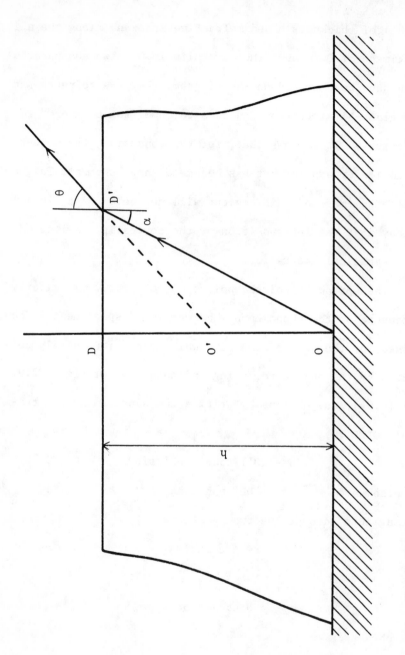

Fig. 3.2.9.(b).

The displacement OO_1 is easily calculated by putting (NA) = 0 in the preceding equation and one finally obtains for the spread $\varepsilon = OO' - OO_1$ the value:

$$\varepsilon = h\{(1/n) - [1 - (NA)^2]^{\frac{1}{2}}[n^2 - (NA)^2]^{-\frac{1}{2}}\} \quad .$$

The spread ε can become quite large; it reaches the value h/n when (NA) = 1.0; then $\theta \to \pi/2$ (grazing emergence) and $O' \to D$. Let us now consider the case of an objective of numerical aperture 0.50. With n = 1.52 and h = 0.18 mm, ε = 12 μm; such a value is much greater than the depth of field of objectives with a numerical aperture of the order of 1/2 and therefore a substantial degradation of the image may be expected. The same expression is found for ε whatever type of illumination (trans- or epi-; bright- or dark-field) may be employed; furthermore, as n is wavelength dependent, ε is also wavelength dependent and therefore with a polychromatic source of light, some chromatic aberrations are to be expected.

Results of a similar nature are obtained in the case of a standard (regular mount) objective: spherical aberration is an increasing function of the numerical aperture of the objective and a decreasing function of its focal length. Substantial progress have been realized in the design of the objectives: for example 10X objectives with a numerical aper-

ture of 0.45 are available which may be used with or without
a coverslip.

A related problem is that of the possible replacement
of glass coverslips by coverslips made of some other material. This substitution presents definite advantages in certain instances, for instance in microchemistry, as in addition to their susceptibility to various types of corrosion
many glasses let alkaline ions leach out. But non-glass materials may have a refractive index significantly different[¶]
from that of crown-glass (n ca. 1.52) which is the accepted
standard for coverslips. For purposes of simplification we
shall proceed as previously, that is to say we shall first
analyze the case when the objective used is of the short-
mount type. As far as axial and paraxial rays are concerned
there is little difficulty: two coverslips made of different
materials will be equivalent if they are such that:

$$h_1(1 - n_1^{-2})/n_1 = h_2(1 - n_2^{-2})/n_2 \qquad .$$

This stems from the fact that for small values of the numerical aperture, the expansion of ε written as:

$$\varepsilon = h\{(1/n) - [1 - (NA)^2]^{\frac{1}{2}}[1 - x^2]^{-\frac{1}{2}}/n\} \quad , \quad x \equiv (NA)/n$$

takes the value $\varepsilon \simeq (NA)^2(1 - n^{-2})/(2n)$

¶   E.g., for cellulose butyrate, n ca. 1.47, while for
cellulose nitrate, n ca. 1.51 and for celluloid n ca. 1.53.

Then the expression $\varepsilon_1 - \varepsilon_2$ which is a measure in the object space of the variation of spherical aberration corresponding to the change of coverslips becomes:

$$\varepsilon_1 - \varepsilon_2 \simeq \{[h_1(1 - n_1^{-2})/n_1] - [h_2(1 - n_2^{-2})/n_2]\}(NA)^2/2 \ .$$

It then appears that, provided that the equivalence relation is satisfied, coverslip substitution does not create any particular problem at low apertures.

When the numerical aperture is not small, one must take extra terms in the development of $\varepsilon_1 - \varepsilon_2$. It is easily found that the equivalence relation does not cancel the coefficient of the term in $(NA)^4$. Thus the latter is in general a non-zero quantity and therefore at medium apertures a substitution of coverslip material is expected to have some influence (adverse or beneficial depending on the values of h_1, h_2, n_1 and n_2) on the quality of the image. What happens at very high numerical apertures is purely of academic interest: when $(NA) \to 1$, the image is always of poor quality (in the particular experimental situation considered), and the fact that $\varepsilon_1 - \varepsilon_2 \to 0$ just indicates that there is no worsening of a bad situation.

For standard objectives, the equivalence relation holds for the axial and paraxial rays; therefore at low apertures the substitution of a glass coverslip by a coverslip made

of some other material does not affect the quality of the
image provided that

$$0.37h_{glass} = h_2(1 - n_2^{-2})/n_2 \quad .$$

At medium and high apertures this substitution generally results in some aberrations.

In many circumstances the introduction of some spherical aberration by the preparation is nearly inevitable: in
addition to the times when the use of a crown-glass coverslip is inadvisable there are instances when it is impracti-
cable to use a coverslip of the proper thickness, for example when examining at high magnification with a dry objective
a series of cemented preparations as it must be recognized
that the systematic replacement of sealed coverslips is - at
best - a tedious operation! Therefore it is again of the utmost importance that the microscopist keep on the alert for
the appearence of some spherical aberration in the images
obtained, and also that he be prepared in critical cases to
carry out the minimization of this aberration. This can be
achieved in two ways. The first procedure requires the use
of a microscope equipped with an adjustable drawtube and is
based on the fact that spherical aberration is a function of
the optical length of the microscope (a general property of
all Seidel aberrations); by gently modifying the value of

this parameter it is possible to bracket the region of minimum spherical aberration. With glass coverslips whose refractive index is of the order of 1.52, an increase of the optical length beyond its nominal value will correct for too thin coverslips and conversely a decrease of the optical length will correct for too thick coverslips (and also for the aberration due to the use of a thin coverslip with a short-mount objective). But note that this manipulation changes the magnification.

When the optical length of the microscope cannot be varied, one can resort to the use of special objectives which include an optical group movable with respect to the front lens by means of an helicoidal ramp controlled by an external knurled ring (correction collar) adjustable by the operator: this device permits one to control to a certain extent the path of the rays within the objective. On the other hand, in the air space between the front lens of the objective and the upper face of the coverslip the path of the rays coming from the object is, in particular, a function of the thickness of the coverslip: it is easily conceivable that things can be managed in such a way that a balance is achieved and that the rays finally converge properly to form a nearly Gaussian image of the object.

The correction collar is generally graduated in terms of coverslip thickness (for n *ca.* 1.52). It is often stated that when the coverslip thickness is known it is sufficient to dial it on the mount; however this is not necessarily the best setting. If the element observed is not in intimate contact with the lower face of the coverslip, the effective coverslip thickness is the sum of the thickness of the coverslip and of the thickness of a certain layer of mounting medium. And in many instances the refractive index of the latter is unknown. Under these circumstances, except when the preparation has been directly mounted on the coverslip (as it is customary for example in tissue culture technique) and not on the slide, it is advisable to determine experimentally, by a bracketing procedure, the best setting of the correction collar. It must be emphasized that, due to the presence of the movable group within the objective, the latter has a varying focal length and that in consequence the magnification slightly changes with the setting of the correction collar.

The case of the phase contrast objectives must be discussed separately as here thre is an additional phenomenon involved. Essentially, in the phase contrast method of illumination one recombines beams of light which have experienced

different fates; the quality of the image depends on the
precision with which the various beams can be manipulated,
and it follows in particular that it is essential for the
success of the method that their proper orientation in the
object space be maintained. A variation of coverslip thickness induces in that space a substantial variation of the
angle between a non-paraxial ray and the optical axis, that
is to say a variation of beam divergence. It is therefore
imperative to use coverslips of the proper thickness, even
at low apertures. But when phase contrast objectives are
used with the brightfield method of illumination the coverslip thickness condition is obviously relaxed, and the results of the general discussion apply.

3.2.10.   Old objectives.

Although they do not look much streamline, old optical
components and in particular old microscope objectives should
not be systematically discarded as upon examination some of
them may be found to be of the highest quality. This is not
surprising because the production of fine optical systems is
largely a matter of craftmanship. As a matter of fact, some
optical components improve with age; this is attributable
to the formation of an halide film on the exposed surfaces:
this film, which sometimes appears as a kind of tarnish, re-

duces unwanted reflections[¶]. On the other hand certain optical glasses undergo a phase transformation and "crystallize".
Objectives in which this phenomenon has occurred, obviously
are unusable.

3.2.11.   Evaluation of objectives.

In addition to the evaluation of the degree of correction of an objective[§], it may occasionally be necessary to
determine the approximate value of some optical parameters.
The working distance of low power objectives can be measured
with the help of dividers: one of the pins should be covered
with thin soft plastic tubing in order to avoid scratching
the front lens. The depth of focus (depth of field) can be
estimated through the use of the graduated micrometric screw
of the fine adjustment: it is obtained as the maximum range
within which a reasonably sharp image of a fine detail can
be obtained.

The optical length for which the objective was calculated can be obtained by measuring, using a monocular instrument equipped with an adjustable drawtube, the distance

---

[¶]   This observation was originally made by H.D. Taylor at
the end of the nineteenth century, on astronomical lenses;
it spurred a great deal of interest in the theoretical study
and the manufacture of anti-reflection coatings.
[§]   Which can be carried out using the techniques described
in 3.1.4.

for which spherical aberration is minimum. The magnification of an objective for a given optical length can obviously be ascertained by the simultaneous use of a calibrated stage micrometer and of a calibrated micrometer eyepiece[¶].

The exact measurement of the most useful parameter of an objective, namely its numerical aperture, is out of question in most laboratories as an apertometer is a rather specialized instrument. It is however possible to arrive at a reasonable estimate of a NA value if a graduated condenser is available: the aperture iris diaphragn is adjusted so that the image of the aperture just covers the back focal plane (examined with the naked eye or better, with a Bertrand system) of the objective. If the condenser has not been calibrated at the factory, this operation will have to be carried out by the microscopist: a strip of adhesive paper is wrapped around the condenser barrel or its mount, and the position of the iris diaphragm tab corresponding to known apertures (using the same technique as above but in reverse, that is to say with objectives whose aperture is known) inscribed. Then the unknown objective is swung into place and its numerical aperture ascertained by interpolation. When a polar-

---

¶   See Chapter 7.

izing microscope is available, the numerical aperture of an
objective can be estimated from the results of the conoscopic
observation of a set of known biaxial crystals.

3.2.12.   Protection of objectives.

Certain manipulations are likely to endanger the objectives if the latter are not given some kind of protection.
The most common causes of damages are heat and the action of
corrosive vapors on the front lens (e.g., hydrofluoric acid),
on the optical cement (e.g., vapors of organic solvents) and
on the barrel (e.g., acid vapors). In such conditions a mono-
objective microscope is to be preferred; alternatively, unnecessary objectives should be removed from the turret and
the corresponding openings closed with a few layers of tape
or, better, by means of threaded plugs. In many instances
too, the interposition between the stage and the microscope
bodytube of a baffle with an opening for the light beam will
reduce the acuteness of the problem: an asbestos sheet, which
is corrosion resistant and also a good thermic insulator, is
very convenient.

1. Heat effects.

They are inevitable in a number of operations, for instance when measuring melting points with a hot stage. Even
when the latter has been designed with care it radiates much

heat; one or a combination of the following means can be used:

1. a reflecting cup, made for instance from heavy duty aluminium (household type), is fixed with rubber bands and plastic tape at the tip of the objective;

2. a jet of air can be directed at the nose of the objective, using a small nozzle made from a piece of brass tubing flattened at its extremity;

3. in extreme cases, a small cooling coil of brass or copper tubing can be closely fitted to the barrel of the objective, and a circulation of cold water established.

2. Corrosive vapors.

For special studies it may be advisable to construct a microchamber connected to an efficient exhaust system; in many cases the latter needs to be no more complicated than a good water aspirator.

The suggestion has been made to enclose the objective in a concentric chamber made of metal and closed by a glass or quartz window (celluloid should be used in the presence of hydrofluoric acid vapors). In fact this system is mecanically complicated because the window must be exactly perpendicular to the optical axis in order to prevent a serious

degradation of the image, and its use is best restricted to
specialized laboratories.

For the microscopist who engages in chemical work occasionally only, the simplest way is probably to fix temporarily a thin coverslip (of glass, celluloid, etc...) to the
front lens of the objective by means of a very small drop of
an immersion fluid such as cedarwood oil. The coverslip,
which acts as a transparent screen, is held in place by
surface tension forces; it is easily positioned perpendicularly to the optical axis as in a number of objectives the
front surface is plane: in the Lister design, for example,
the front lens is constituted by a plano-concave meniscus
cemented to a biconvex lens.

3.3.   EYEPIECES.

3.3.1.   Functions.

The main function of an eyepiece is to project an image
of the primary image, that is to say of the real image furnished by the objective (and eventually the intermediate
stage), on a physical device (e.g., a photographic emulsion)
or on a physiological system, the eye. It should be first
noted that in some instances one may dispense entirely with
the eyepiece: this is for example the case in low or medium
power photomicrography when an image of the object under in-

vestigation is projected on the film by the sole objective
and in conoscopic observation by the Lassaulx method. In most
cases however an eyepiece is used to impart some additional
magnification to the primary image, and often to correct this
primary image (compensating eyepieces) or to modify it and
extract some specific information (e.g., spectroscopic eyepieces).

It is of the utmost importance to realize that the ulti-
mate of the image projected by an eyepiece determines how
this projection is carried out. By far the simplest case is
that of the exploitation of the image by a purely physical
device; then, a real image must be projected by the eyepiece
and therefore the whole microscope must be adjusted (focused)
so that the primary image is formed outside the first focal
plane of the eyepiece. In visual observations the situation
is more complex as it depends in particular on the way in
which the eye of the observer is accommodated. When the eye
is accommodated at infinity, which is the recommended procedure as it makes for less fatigue, the eyepiece must project
an image at infinity and consequently the primary image must
be located in the first focal plane of the eyepiece.

In the same way, the exit pupil of the whole microscope,
whose size and height above the eyelens are dependent on the

eyepieces, must be matched with the entrance pupil of the system on which the image is projected: a large entrance pupil
means a wide angular aperture (a very desirable feature) but
on the other hand aberrations increase when the aperture increases, and therefore there exists a limited optimal range
for the size of the eyepiece exit pupil. In the case of the
eye for example, aberrations become marked when the pupil
diameter is greater than 5 mm; thus eyepieces specifically
intended for visual observations are best constructed in such
a way that the diameter of the exit pupil lies within the
range 4 - 5 mm.

3.3.2.   Classification of eyepieces.

Eyepieces are classified according to:

1. their sign: in a positive eyepiece the first focal
    plane of the system is outside the eyepiece, while
    it is located inside in the case of a negative eye-
    piece;

2. the name of the designer or of the design: one
    speaks for example of Huyghens, Ramsden, orthoscopic
    eyepieces;

3. the type of corrections varried out: e.g., achromat-
    ic, flat field eyepieces;

4. their intended association with certain objectives:

this is in particular the case of the compensating eyepieces associated with semi-apochromats;

5. special optical characteristics: e.g., wide field eyepieces and erecting eyepieces;

6. their intended use: e.g., projection, demonstration, goniometer, micrometer[¶] eyepieces;

7. the type of accessory with which they are used or within which they are incorporated. There may be some ambiguity in that case: while a spectroscopic eyepiece definitely is a microspectroscope or a microspectrograph, a photographic eyepiece may be either a projection eyepiece specially designed for use in conjunction with a camera or an integrated device including a projection eyepiece and a photographic chamber.

3.3.3. Magnification.

The magnification obtained with a given eyepiece depends on the way in which the eye is accommodated. In the case when the eye is accommodated at 25 cm (this length is the least distance of distinct vision for the normal, or emmetropic, eye and is taken as a standard value in physiological optics),

---

¶   Or micrometric.

the linear magnification is

$$G_{25} = 1 + 250/|f|$$

where f is the focal length, expressed in mm, of the eyepiece. When the eye is accommodated at infinity, as it is the
standard (and advocated) practice in microscopy, the magnification becomes:

$$G = 250/|f| \qquad .$$

The latter value is taken as the characteristic magnification
of an eyepiece and is at times referred to as the magnification number.

The dependence of the magnification on the focal length
makes it necessary to investigate this class of eyepieces
whose distance between the field lens and the eyelens is variable; this group includes in particular the eyepieces whose
eyelens is focusable on a crossline reticle or a micrometric
scale, and the projection eyepieces. The variation of magnification associated with the adjustment of such eyepieces is
easily estimated with the help of the classical Gaussian
formula which gives the focal length F of a system composed
of two thin lenses of focal length f_1 and f_2 respectively,
separated by a distance h:

$$1/F = 1/f_1 + 1/f_2 - h/(f_1 f_2) \qquad .$$

Let us consider the case of a Huyghens eyepiece for which

$f_1 = 3f$, $f_2 = f$, $h = 2f$. We have:

$$|\Delta G/G| = |\Delta(1/F)/(1/F)|$$

$$= |\Delta h/(2f)|$$.

We now select the value $\Delta h = 1$ mm. This variation of the separation distance will induce a variation of the magnification smaller than or equal to 1% if $F \geq 75$ mm, $i.e.$, for eyepieces whose nominal magnification is of the order of 3 and smaller: many projection eyepieces belong to this group. For the same small displacement of the eyelens with respect to the field lens, the variation of the magnification will be less than 3% if $F \geq 25$ mm, that is to say for eyepieces whose nominal magnification is no greater than 10. As a matter of fact the classical eyepieces fitted with crossline reticles or micrometric eyepieces are 10X reticles: the effect of a small displacement of the eyelens is large enough to be noticeable but not so large as to be bothersome for qualitative visual observations.

3.3.4.   Field.

The field of all eyepieces is intentionally reduced by means of a stop (field stop) in order to decrease a number of aberrations. The ocular field is related to the object field, that is to say to the field encompassed by the objective, as:

object field = eyepiece field ÷ effective
                                        magnification ,

where the effective magnification is the product of the magnification of the objective by the magnification of the intermediate stage if one is present. This relation is in particular useful for estimation purposes when only the nominal
magnification (as opposed to the actual magnification which
depends upon the optical length of the microscope) of the
objective is known.

3.3.5. Markings on eyepieces.

Markings on eyepieces are few and are usually confined
to the eyelens face mount, although in certain cases some indications are engraved on the barrel: this occurs for instance with projection eyepieces which already bear a scale
(projection distances) on the eyelens mount.

Some eyepieces are characterized by a number only; more
frequently modern eyepieces bear engraved their nominal magnification for a standard eye accommodated at infinity. A
Huyghenian ocular may bear the letter "H", generally engaged
in a combination such as "H8X"; a compensating eyepiece generally bears a notation such as "comp" or a plain "C" (e.g.,
in "C8X"). Bertrand and projection oculars are plainly marked
as such; this is often the case too with micrometer eyepieces.

Eyepieces well corrected for curvature of field are often designated with somewhat fancy names which generally in-

clude the words "plan" or "peri" (e.g., hyperplan, periscop-

ic). They are mainly used in photomicrography but note that,

as a rule, they are but poorly achromatized.

3.3.6. Negative eyepieces.

The prototype of all negative eyepieces is the original

Huyghens design: both lenses are plano-convex, the convex

side facing the objective, and the system is of formula[¶]

3 2 1.

Huyghens eyepieces present many advantages:

1. they give a bright image;

2. their field is rather large;

3. they have little chromatic aberration (specially

   lateral chromatic aberration);

4. internal reflections are negligible.

But they are restricted to low and medium powers as aberra-

tions, both geometrical and chromatic, rapidly increase with

the magnification. Also the eye relief, that is to say the

distance of the apex of the eyelens to the Ramsden circle,

is somewhat small, so that a 15X Huyghens eyepiece is already

difficult to use for prolonged visual observations.

[¶]   The formula of a (centered) two-lens system is an order-
ed sequence of numbers proportional (from left to right) to
the focal length of the front lens, the separation distance
and the focal length of the rear lens respectively.

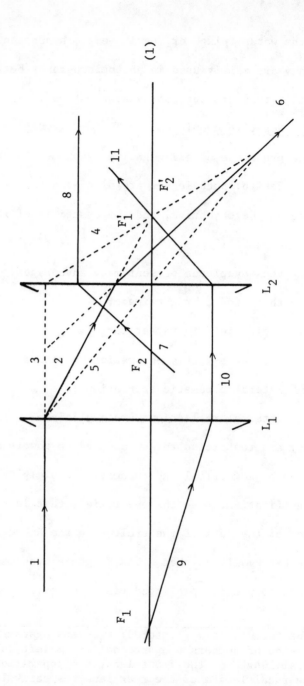

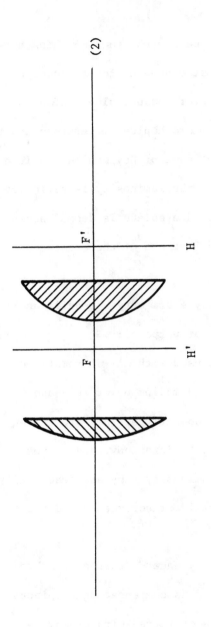

Fig. 3.3.6. Huyghens eyepiece (formula 3 2 1). L_1: field-lens with its foci F_1 and F_1'; L_2: eyelens with its foci F_2 and F_2'. In (1), some important rays have been traced; note that the two points F_1 and F_1' (=F_2') are conjuguate. In (2), some cardinal elements of the system are shown on the same scale: the two principal planes H and H' and the two foci F and F'; note that F is identical with F_2.

Additional corrections can be brought to a Huyghens eyepiece by substituting an achromatic doublet for at least one
of the plano-convex lenses. However the usefulness of such
negative systems, known in the microscopical literature under
the generic (and most confusing) name of Huyghenian eyepieces,
is questionable as in particular the luminosity is seriously
decreased. When a highly corrected eyepiece is deemed necessary, it may be preferable to select one of the orthoscopic
design.

Eyepieces of the Huyghens type are very versatile and
have proved useful for a number of purposes:

1. as achromats, they are used with all achromatic ob-
    jectives and with low and medium power semi-apochro-
    matic and apochromatic objectives. They are current-
    ly used in phase contrast microscopy, and because of
    their high intrinsic luminosity they are practically
    indispensable in darkfield, fluorescence and polar-
    isation microscopy;

2. as over- (and more rarely under-) corrected achro-
    mats, they are much used as compensating eyepieces;

3. as micrometer eyepieces or fitted with crossline
    reticles. In that case the eyelens is independently
    focusable on the plane of the diaphragm (stop) lo-

cated very close to or in the second focal plane
[plane H' in Fig. 3.3.6., (2)].

3.3.7.   Positive eyepieces.

There is on the market a large variety of positive eyepieces, as they are extensively used in a number of optical
instruments. But a few types only of those are of use in microscopy, and there is unfortunately some confusion in the
microscopical literature as at one time it was customary to
label any positive eyepiece as a "Ramsden eyepiece".

The original Ramsden eyepiece which can be taken as the
prototype of the positive eyepiece is a two-lens system of
formula 1 1 1 constituted by two plano-convex lenses whose
convex sides face each other (see Fig. 3.3.7.). Its main advantages are:

1.   to perform well at relatively high magnifications.
      As a matter of fact, it is not well suited for work
      at low magnification;

2.   to have little longitudinal chromatic aberration
      (generally much less than the corresponding Huyghens
      eyepiece);

but:

1.   as many positive eyepieces which include convex sur-
      faces facing each other, it is prone to internal re-

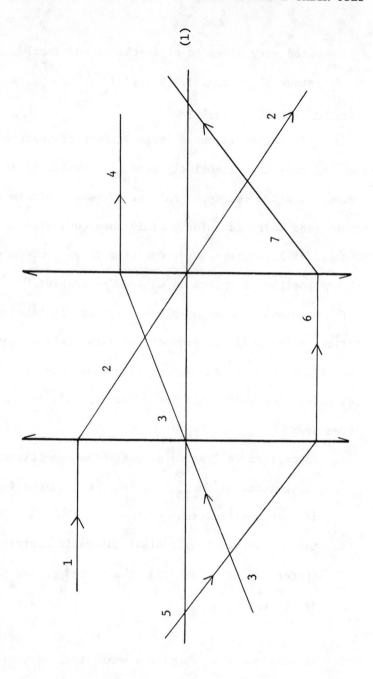

(1)

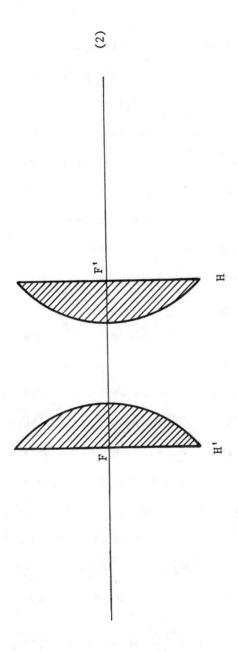

Fig. 3.3.7. Ramsden eyepiece (formula 1 1 1). L_1: field lens with its foci F_1 and F_1'; L_2: eyelens with its foci F_2 and F_2'. In (1), some important rays have been traced; note that the two points F_1 and F_2' are conjuguate. In (2), some cardinal elements of the system are shown: the two principal planes H and H', and the two foci F and F'.

flections can be decreased by the judicious use of
stops and by coating the critical interfaces with
suitable layers (anti-reflection films);

2. any defect of the field lens (e.g., a scratch) or
a speck of dust on the outside glass surface will
be imaged by the eyelens[¶];

3. it performs best when the entering cone of rays is
relatively steep: this is a situation commonly encountered in reflecting telescopes, but not in microscopes;

4. it has more lateral chromatic aberration than the
corresponding eyepiece.

Furthermore the relatively small eye relief of a Ramsden
eyepiece makes it unsuitable for prolonged visual observations when its magnification number is greater than 15. Under
these circumstances, and in spite of its past popularity, the
Ramsden eyepiece cannot be recommended for routine use in microscopy; in those cases when it is adequate, its magnification is best restricted to the range 10 - 15X.

At the present time, the best positive eyepieces are

¶   This disadvantage of the original Ramsden eyepiece is
easily corrected by modifying slightly the separation distance, that is to say by constructing an eyepiece of formula
[1,1+ε,1] where ε is small.

probably those of the orthoscopic[¶] design which has been o-

riginally proposed by Abbe: the eyelens is similar to that

of a Ramsden eyepiece, but the field lens is an achromatic

triplet. Formerly a symmetrical triplet (one internal menis-

cus cemented to two identical convex lenses) was used. Nowa-

days the Hahn design seems preferable: the front lens of the

triplet is ground as a plano-convex lens. Orthoscopic systems

can be highly corrected, and their eyepoint is relatively

high; unfortunately they are not very luminous, but in the

range 16 - 25X they stand unrivalled, and can be obtained

either as achromats or as over-corrected achromats to be used

as compensating eyepieces.

Wide field eyepieces generally include a third lens,

single or compound, which acts as an extra field lens. They

often suffer from astigmatism and distortion.

3.3.8.   Compensating eyepieces.

Strictly speaking, compensating eyepieces are designed

to correct the residual aberrations of certain well deter-

mined objectives. Experimentally however, and except for very

critical observations, compensating eyepieces calculated for

use with contemporary semi-apochromats and produced by dif-

[¶]   Although this word was specifically coined to describe
an optical system, it has sometimes been used (regrettably)
to indicate a fine degree of performance.

ferent manufacturers can often be interchanged. The use of
such compensating eyepieces with achromatic objectives is
however more restricted as, in spite of affirmations to the
contrary[¶], they are not recommended for use in conjunction
with low power achromats: this combination has, at times,
been reported as presenting marked chromatic aberrations.
Therefore caution should be exercized and the particular combination compensating eyepiece - achromat should be tested
for aberrations.

It has often been stated that, when examined with an extended source of light, a compensating eyepiece shows a yellow fringe at the border of the field. This is indeed the
case with eyepieces designed to compensate the residual aberrations of many semi-apochromats, but no general statement
can really be made as such a yellow fringe is not observed
with particular eyepieces calculated to compensate the residual aberrations of certain highly achromatized objectives.

3.3.9.   Bertrand oculars.

These oculars, also called telescopic eyepieces, are
non-erecting short-focus telescopes which are mainly used to
examine the back focal plane of an objective when adjusting

¶   Even in one manual of laboratory techniques!

a phase contrast microscope. They constitute a modern version
of the classical Bertrand system which has proved most useful
in optical crystallography: (with the polarizer and the analyzer inserted in crossed positions) an adequate convergent
lens is interposed between the ocular and the objective, thus
transforming the regular eyepiece into a small telescope
through which the observer can study interference figures in
the back focal plane of the objective (conoscopic observation).

A Bertrand ocular gives an enlarged image of the back
focal plane of the objective, but this image may suffer from
some lack of sharpness. For other means of observing this
focal plane, see 6.2.2.

3.4.   CONDENSERS AND ILLUMINATORS.

3.4.1.   Functions.

The terms "condenser" and "illuminator" have often been
used indiscriminately; this is largely a problem of semantics
and for purposes of simplification we shall use here the word
"condenser" in the strictest sense of the term, that is to
say to designate a centered optical system which essentially
includes lenses and whose function is to provide a light beam
adequate for a given type of illumination or to modify a
light beam so that it becomes adequate for this purpose,

while the word "illuminator" will be exclusively reserved for these devices which do not include lenses and are generally used in conjunction with a lensatic system.

Originally the sole function of a condenser was to concentrate (= to condense) on the preparation the light from a source. But as microscopists began to strive for smaller and smaller limits of resolution, the importance of the convergence of the illuminating beam was slowly recognized. Other roles were also progressively devolved to the condenser: to control the shape of the illuminating beam (e.g., in dark-field microscopy), to smooth out small variations of the brilliance of the source (in Köhler's method of illumination), to project a scale in the plane of the preparation for micrometry, etc... And many a specialized method of microscopy depends in an essential way upon certain characteristics of the condenser: for example, a very high convergence is desirable for the study of interference figures[¶], while Zernike phase contrast illumination method requires the precise imaging of the condenser aperture diaphragm on a certain region of the back focal plane of the objective and therefore the aplanatism of the condenser becomes critical.

[¶]   Generally with polarized light; see any manual of optical crystallography.

3.4.2.   Classification of condensers.

Condensers are usually classified according to:

1. the eventual corrections brought to the system. Customarily one distinguishes Abbe, aplanatic, achromatic and planachromatic condensers;

2. the presence of special optical features (e.g., reflecting, cardioid condensers);

3. their intended use (e.g., darkfield condensers, phase contrast condensers[¶]);

4. the nature of the refracting substance used if it is not one of the regular optical glasses (e.g., quartz condensers);

5. (rarely) their numerical aperture. A high aperture condenser may be taken as having a numerical aperture of the order of 1.40;

6. the presence of particular mecanical features (e.g., swing-out top lens).

3.4.3.   Correction of condenser aberrations.

The classical Abbe condenser presents many advantages: it is relatively cheap to produce and can impart a high convergence to the illuminating beam (in terms of numerical ap-

---

[¶]   Often abbreviated as "phase condensers".

erture the practical upper limit attainable with this design
is of the order of 1.30). It has proved very convenient for
routine work in brightfield microscopy with transmitted light
and for polarization work. Also, as the losses of light of
this device are small it has been for a long time a favorite
for fluorescence work and because it can be manufactured with
a relatively large working distance, it is much used in chemical microscopy. But it has two serious optical defects: it
is far from being achromatic and is afflicted with an evident
lack of aplanatism. However, as it was mentioned earlier[¶],
its first deficiency may at times be turned into an advantage.

The main theoretical result in this domain is an optics
theorem established by Zernike[§]: the presence of aberrations
in the condenser does not entail a loss of resolving power
for the objective. This statement, which is of an extreme importance, must be carefully interpreted in an experimental
context.

Let us note in the first place that strictly speaking
Zernike's proposition, as well as the subsequent analyses by
Born and others, refers to quasi-monochromatic light; but the
extension to a polychromatic beam is fairly immediate in the

[¶]   See Chapter 1.
[§]   And also, at a later date, by Born and Wolf.

case of an objective used in bright- or dark-field microscopy. Now the Abbe condenser, in spite of its optical flaws, has permitted painstaking histologists and cytologists to use the full resolving power of many an objective. This can be taken as an empirical confirmation of the (extended) Zernike conclusion.

On the other hand the role of a condenser is not limited to illuminating the preparation. A good aplanatism is obviously desirable when one wishes to project a scale in the plane of the preparation for micrometric work. In phase contrast microscopy, aplanatism and eventually a good achromatism, are characteristics much sought for, as a strict control of the illumination beam is essential for attaining high quality images; etc...

3.4.3.   Vertical illuminators.

Vertical illuminators are devices interposed (directly or through a revolving nosepiece) between the microscope bodytube and an objective[¶], and which essentially include a reflecting element which deflects - through the objective and towards the preparation - a light beam supplied either by an external source and a condenser or by built-in illumination

---

[¶]   Generally of the short-mount type.

systems.

The most classical vertical illuminator is of the Beck[¶]
design: a thin glass plate (e.g., a coverslip) is set at near
45° to the optical axis of the microscope and approximately
centered (see Fig. 3.4.4.). Thus part of the light is collected by the objective and used for the illumination of the
preparation. In the original design of this accessory, the
glass plate is mounted on a short rod which can be rotated
by means of an external knob: the angle of the plate with the
optical axis of the microscope is varied until optimal conditions of illumination (luminosity and contrast of the image)
are encountered.

There obviously exists a great number of variations
around this basic pattern. Note in particular that if an optical shop is available, a very convenient illuminator can be
made from a half-reflecting plate (of thickness 1 - 2 mm)
ground to size[§] and glued at 45° in a barrel of short length
fitted with two RMS threads (one male and one female).

The main advantage of this type of vertical illuminator
is that the aperture of the objective is fully utilized; but

[¶]   At times it is referred to as a transparent reflector
vertical illuminator.
[§]   This operation must be performed before evaporating on
the plate a suitable layer.

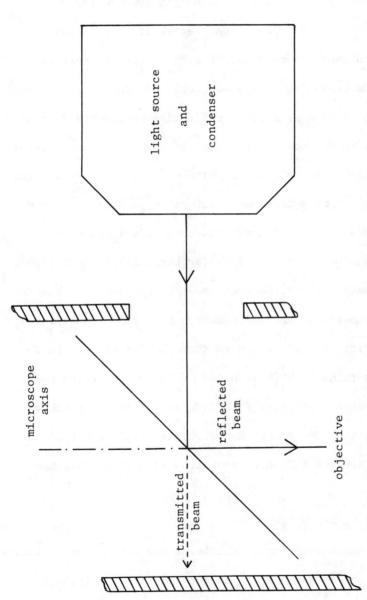

Fig. 3.4.4. Diagrammatic cross-section of a Beck vertical illuminator. The transmitted beam is absorbed, for example by a layer of black paint.

the light efficiency of these systems is low and images of
high contrast cannot be obtained. Better light efficiency and
stronger contrast are made possible by the use of vertical
illuminators providing an asymmetrical illumination. In these
devices use is made of totally reflecting components disposed
laterally with respect to the optical axis of the microscope:
either a tiny total reflection prism or a small mirror. These
systems present many disadvantages, as in particular they do
not allow one to study the variations of the image when one
passes from a purely symmetrical (brightfield) illumination
to an asymmetrical one. However one may vary somewhat the degree of asymmetry of the illumination by displacing radially
the reflecting element. But note that the latter must be located as near the back focal plane of the objective as possible so that no appreciable shadow is cast in the plane of
the preparation; the smooth and accurate displacement of the
reflecting element in these conditions is a trying mecanical
problem.

3.5.   MIRRORS AND PRISMS.

Most microscopes include at least one reflecting device,
often a plano-concave mirror[¶] used to deviate the light from

---

[¶]   This accessory is classically referred to as "the mirror".

the source towards the condenser; a plane mirror is found in many built-in illumination systems, another one in trinocular heads deflects the beam exiting from the objective towards the eyepieces, etc... while prisms are traditionally employed in binocular attachments.

First surface mirrors are exclusively used. Silvered first surfaces are extremely fragile and for this reason are not very suitable in equipments which will see much field work; also, the reflectance of silver which is nearly constant over the visible part of the spectrum sharply drops for wavelengths below 400 nm[¶] so that silvered reflecting surfaces are practically useless for wavelengths smaller than 350 nm. On the other hand, in contrast to aluminizing procedures, the silvering of glass surfaces does not require any specialized equipment and therefore is sometimes used for the fabrication in the laboratory of special reflectors: it is carried out by reduction (e.g., with lactose) of a silver salt, generally the nitrate.

A total reflection prism may often be substituted for a plane mirror: it is less fragile than a first surface mirror

---

[¶]   This property is occasionally used for filtering out visible light: a quartz lens coated with silver will let pass only the UV rays.

and, when it is made of quartz, performs well in the UV region. Such a system has been used for reflecting the beam
from an external source towards the condenser; note however
that when the microscope used is of the horseshoe type, it
may be advantageous to dispense entirely with the reflecting
device and to align the source on the microscope axis[¶].

3.6.   INTERMEDIATE STAGE.

In some research microscopes the beam exiting from the
objective passes through a low magnification (a typical value
is 1.3X) intermediate stage[§] before reaching the eyepieces.
The main advantage to be derived from the presence of this
auxiliary component resides not so much in the additional
magnification it imparts[†] to the image as in the decrease it
induces in the convergence of the beam. This is a most valuable property in binocular microscopes which heavily rely on
prisms to split the main beam and to deflect the two daughter
beams towards the eyepieces: spurious polarization phenomena
are likely to occur if the rays are too much inclined on the

¶   See 4.6.3.
§   It is probably because relatively few microscopes are
equipped with this device that there does not seem to exist
a standard name for it. The expression "intermediate stage"
which we use here should not be considered as something more
than an expedient.
†   It is for a microscope the equivalent of a Barlow lens
for a telescope.

optical axis. This reduction of the beam convergence also fa-
cilitates high magnification photomicrography.

The presence of an intermediate stage does not modify in
any way the standard optical procedures and according to the
situation one will find it advantageous to link the device
either with the objective or with the eyepiece. We shall
speak for example of the effective magnification of the objective as the product of the magnification of the objective
by the magnification of the intermediate stage (generally engraved upon its mount). In these procedures in which one observes the back focal plane of the objective, as for instance
when adjusting a phase contrast illumination system or in
conoscopic observations with polarized light, it may be easier to consider the intermediate stage as an extension of the
eyepiece(s).

The involvement of a third component in the formation of
the image obviously raises the question of an eventual modification of the residual aberrations of the (partial) system
objective - eyepiece. In most cases this effect is extremely
slight, if noticeable at all; this is not very surprising as
it is relatively easy to correct an optical system of low
magnification. However it would be prudent to keep in mind
the possibility of an influence of the intermediate stage on

the quality of the image, specially when working near the extremities of the visible spectrum.

3.7. THE CLEANING OF OPTICAL ELEMENTS.

It was formerly admitted that the optical elements of a microscope did not require much cleaning, but this attitude must be reconsidered because the complex optical and mecanical design of a modern research microscope has for consequence the creation of a number of dust traps, e.g., the upper face of the prisms in a binocular instrument: they are exposed to the room atmosphere each time one changes the eyepieces, and therefore are bound to collect some dirt. Furthermore most microscopes are far from being tightly closed systems and pollution deposits easily find their way to many air-glass interfaces: this problem can obviously be minimized by keeping an instrument covered with a plastic sheet when not in use, but nevertheless it remains ever present.

The difficulties one has to face in the cleaning of optical components are of two orders, mecanical and chemical:

1. as a rule, optical glasses are softer than regular glasses and are easily scratched; anti-reflection coatings too are fragile. Therefore all cleaning manipulations should be carried out in a reasonably dust-free environment and all the materials which

come in contact with optical glasses should be free from abrasive particles. Good quality lens paper can be bought from reputable manufacturers or supply houses, but care must be taken that it does not become contaminated in the laboratory; a suitable procedure is to separate lens paper booklets into small lots which are individually preserved in airtight plastic bags. If no lens paper is available, laboratory-type tissue paper may prove suitable: it should be checked for the absence of silica particles, for instance by examination between crossed polars. Finally, repeatedly washed and boiled pieces of natural fiber cloth (with the exception of cotton which gives away many lints) may be used. Again, the absence of abrasive particles should be microscopically checked;

2. the cements used in optics are fragile, in the sense that they are far from inert when exposed to many organic vapors. It must be emphasized that the softening of the cement may be sufficient to allow a slight displacement of a lens or of a group of lenses: this may prove to be the ruin of an objective! This explains why organic solvents should be used

very sparingly only, and only for very short times;
this rule applies even in those situations where the
use of a specified solvent is permissible, e.g., for
the cleaning with xylol of the front lens of an oil-
immersion objective. Xylol remains the standard solvent for this kind of operations; in stubborn cases
ethanol can be used, but it is risky.

Often, breathing on the glass surface followed by a
gentle rubbing with lens paper will clear the problem. If it
fails, a treatment with saliva may be tried: it possesses
slight detergent properties and possibly its efficiency as a
cleaner is due to its enzymatic properties. Saliva can definitely be recommended for removing tarnish on lenses and
mirrors. Note that water is generally useless for removing
contaminants on glass (with the exception of water-immersion
objectives contaminated by water-soluble compounds). Tap
water is worse, as it leaves deposits of its own on the glass
surface. If everything fails, it is advisable, before trying
to use powerful solvents, to contact the manufacturer (he
should know the exact nature of the cement used!) or to ask
the advice of a competent optical engineer: many more eyepieces and objectives have probably been ruined through unwise cleaning than direct misuse.

C H A P T E R   4

ILLUMINATION OF THE PREPARATION

4.1.   GENERALITIES.

The worth of an adequate illumination of the preparation
was already familiar to early microscopists. There is some
evidence that with the simple microscopes (single lens magni-
fiers) he was constructing, Anton van Leeuwenhoek employed an
illumination method akin to a darkfield technique; anyway he
must have been quite conscious of the importance of illumination procedures in microscopical observations, as he constantly refused to disclose the method he was using! Another
founding father of microscopy, Robert Hooke, was currently
emplying a condenser of the bull's eye type as it is clear
from one of the illustrations of his *Micrographia*.

The quality of the microscopical image strongly depends
upon both the proper selection and the judicious adjustment
of the illuminating system. This is not only a matter of op-
timization, of obtaining the best from one's equipment; it
must be emphasized that a low grade illumination can be

responsible for a number of artifacts and of spurious images.
This is easily demonstrated by tampering with a Köhlerian
system of illumination: if an uneven image of the filament is
projected in the plane of the iris diaphragm of the condenser
it functions as a slit diaphragm and diffraction fringes will
often be observed. The experiment is best made using a binocular microscope equipped for phase contrast work: a standard observation eyepiece is placed in one of the tubes and a
Bertrand ocular in the other. A rotation of an uneven source,
easily followed with the Bertrand system as the annulus of
the phase plate is not evenly filled with light, will result
in images (observed with the standard eyepiece) which have
little in common: at times one can even record an apparent
reversal of contrast!

4.2.   SYMMETRICAL AND ASYMMETRICAL ILLUMINATION.

It is perhaps because of the cylindrical symmetry embod-
ied in a microscope that many investigators exclusively use
a nearly symmetrical illumination system and do not avail
themselves of the possibility of adding some shading to the
image (and also of increasing the power of resolution of the
objective) by resorting to an oblique illumination procedure;
this technique has practically fallen in oblivion, except in
optical crystallography, and in fact a systematic discussion

of illumination techniques in terms of all the possible sym-

metries does not seem to have ever been made.

A *priori* the basic cylindrical symmetry of the micro-

scope can be broken in three different ways, or in other

words three different possible asymmetries of the illumina-

tion must be considered:

1. geometrical asymmetry: the axis of the illuminating

   beam does not coincide with the axis of the body-

   tube;

2. photometric asymmetry: the distribution of the light

   intensity in a cross-section of the illuminating

   beam is not homogeneous;

3. chromatic asymmetry: the illuminating beam is not

   photometrically homogeneous for certain wavelengths.

Let us note also that the plane of the preparation con-

stitutes a natural plane of symmetry for illumination methods;

it follows that, from the point of view adopted here, there

is no need to establish a distinction between epi- and trans-

illumination systems.

Conditions of geometrical symmetry can be recognized by

means of a simple and very sensitive procedure described by

Gage: it is based on the fact that small globules are prac-

tically spherical and can be made to perform as tiny lenses.

A suspension of oil globules, or maybe more conveniently of air bubbles[¶], in water is examined through the microscope: when the illumination is geometrically symmetrical, the image is circularly symmetrical. The air bubble test of Gage can be carried out with objectives of any power, but it is not well suited for work with oil-immersion objectives. In that case one may prefer to observe diffraction fringes around very small objects (e.g., pigment granules, diatoms): if the illumination is symmetrical, the fringes will substantially have the same width. Photometric and chromatic symmetries are easily recognized by inspection of the field; while the latter is most easily achieved by interposition of a filter, the former (evenness of the field) may prove more difficult to realize[§].

Pure geometrical asymmetry is easily achieved, e.g., with the help of a plane mirror deflecting a beam of parallel light (photometrically and chromatically homogeneous) toward the preparation. In many instances the illuminating beam is endowed with both geometrical and photometric asymmetries:

[¶]   A preparation of air bubbles may be obtained by beating some air, with a small glass rod or a dissecting needle, in a proteinaceous solution (e.g., a dilute suspension of egg white or gelatin, a drop of saliva).
[§]   See section 4.4.

this is often the case when illuminating the preparation with the help of a concave mirror, or when blocking part of the aperture of the main condenser. The latter technique known in optical crystallography under the names of oblique illumination method or half-shadow method is much used to compare the refractive index of a solid particle[¶] with that of its surrounding liquid medium: the illuminating beam exiting from the main condenser is rendered oblique by screening some of the light entering this condenser. The aspect of the field is depicted in Fig. 4.2.

The success of the method depends on the realization of a certain number of conditions: the most important one is that the main condenser be properly focused, as the phenomenon is reversed when this condenser is focused under the element observed (for this reason, some microscopists automatically rack up the condenser before proceeding with the test). The phenomenon is sharper when the convergence of the illuminating beam is kept low and when the numerical aperture is

---

¶   With isotropic substances, the test may be carried out with natural light; with anisotropic substances, one makes use of polarized light: between crossed polars, the selected crystal is brought to the center of the field and the stage is rotated until the crystal is in a position of extinction. Then one of the polars is removed and the oblique illumination test carried out.

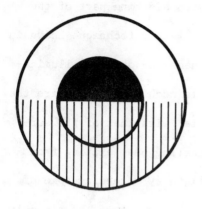

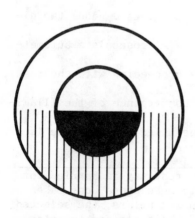

Fig. 4.2. Diagrammatic aspect of the field in the half-shadow method. On the left, the refractive index of the particle is higher than that of the immersion medium; on the right, the refractive index of the particle is lower than that of the immersion medium.

rather small, and therefore the use of objectives which incorporate an iris diaphragm in their back focal plane has
been recommended. Two classical variants of this system permit one to increase the sensitivity of the test. They both
make use of an auxiliary light baffle located in the back focal plane of the objective: in Wright's method, this stop is
on the opposite side of the condenser screen, while in Saylor's method the two screens are positioned on the same side.

In optical crystallography the test is generally carried
out by partially blocking the beam entering the main condenser with the fingers or a small plate (cardboard card, sheet
of opaque plastic, etc...). This procedure is obviously unsuitable for prolonged observation or photomicrography. In
such cases the use of an auxiliary eccentric iris diaphragm
often proves convenient: the iris diaphragm can be moved laterally, in a plane perpendicular to the optical axis, by
means of a rack and pinion drive. This device which is interposed between the auxiliary and the main condensers, is
mounted on a cylindrical sleeve which can be rotated around
the main condenser axis.

In the preceding discussion of asymmetrical illumination
the convergence of the illuminating beam and its shape did

not appear[¶]; it immediately follows in particular that this
discussion is equally valid for bright- and dark-field microscopy, with a plain cone of light or with annular illumination, etc... This suggests the possibility of extending the
use of asymmetric illumination to domains where traditionally
symmetric illumination only has been used. In fact a few pre-
cedents are known, as in addition to the oblique illumination
technique (in its classical version and in the Wright variant) there exist two illumination methods which essentially
rely on at least one of the asymmetries aforementioned: i)
Rheinberg illumination and especially its variants[§] make use
of a chromatic asymmetry, and ii) in the Saylor modification
of the half-shadow method, the two screens can be adjusted
so that the conditions of a geometricallly asymmetrical darkfield illumination are realized. Undoubtedly, much remains
to be done in this whole field.

¶   But they may appear in technical expressions which refer
to often used physical configurations. In particular, symmetrical illumination with a plain cone of light of low convergence, which is much used (for example for the measurement
of the refractive indices of a crystal by the Becke line
method), is often referred to as "axial illumination". It has
often been advised to realize axial illumination conditions
by both closing the aperture iris diaphragm of the main condenser and racking down the latter; the last manipulation is
of dubious value.
§   See 4.5.6.

4.3.   LIGHT: SOURCES, COLOR AND INTENSITY.

4.3.1.   Light sources.

For the microscopist the most important characteristic of a light source are its area, evenness, color [see Table 4.3.1.(a)] and intensity. The size and evenness of a light source are decisive factors in the selection of an adequate system of illumination and their influence is discussed in section 4.4.; the intensity of a light source becomes critical in these procedures where a very small fraction only of the light reaching the preparation is collected by the objective, as it is the case in immersion ultramicroscopy, or when one has to isolate a very narrow band of the spectrum for instance by means of filters.

The classical sources of light in microscopy are white light sources, generally of the incandescent type, whose spectral distribution is smooth (this is not for instance the case with fluorescent lamps whose spectrum contains very strong emission lines). Physically speaking, white light sources can be characterized by their color temperature, that is to say the temperature of a black body with the same spectral distribution. The knowledge of this parameter [for representative values, see Table 4.3.1.(b)] is of interest in photomicrography as this permits one to predict to some ex-

TABLE 4.3.1.(a).

SUBDIVISIONS OF THE VISIBLE SPECTRUM.

Note that the terminology used is somewhat arbitrary; for example, in the graphic arts the separation of the blue and violet regions is set at 450 nm (indigo).

Wavelength (in nm)	color
400 ---	
	violet
435 ---	
	blue
500 (turquoise) ---	
	green
570 (strontium lemon-yellow) --------------------------------	
	yellow
590 (chrome orange) ---	
	orange
610 (vermilion) ---	
	red
700 ---	

tent what types of filter should be used to match the color
temperature of the source and that of the film but it must be
emphasized that one cannot expect much more than a crude indication: the candid use of the same filters as those recommended in common photography is most likely to result in an
image of poor quality as often the preparation deeply changes
the spectral distribution of the light, for example by select-
ively absorbing certain bands.

Numerous generations of microscopists have relied on the
diffuse light from the sky, whose use is mostly restricted to
brightfield microscopy with trans-illumination: in the same
way old flame sources such as the liquid paraffin lamp (or
even a plain candle!) may prove useful for field work. A special place must be attributed to the sun as it is still an
ideal source for resolving very fine colloids with a Zsigmondy immersion ultramicroscope: it is however very difficult
- as well as dangerous - to use and in practice an heliostat
is required, so that the implementation of this technique is
reserved to a few highly specialized laboratories.

The carbon arc, much popular in microscopy laboratories
in times now past, has today been superseded by modern electric bulbs whose light yield has been steadily increasing
during the last decades; it does not follow however that it

TABLE 4.3.1.(b).

APPROXIMATE COLOR TEMPERATURE OF VARIOUS LIGHT SOURCES.

The color temperature can be raised by means of a blue filter or decreased with a
yellow or red filter.

°K	Light source
6000	blue flashbulb
5500	"daylight" = sunny day skylight around midday,
	xenon arc
3800	clear flashbulb
3200 – 2600	low-voltage microscope lamp

should be completely discarded, because techniques such as
emission histospectrography are still clearly susceptible of
many applications! Simple light sources include the classical
40 - 60W opal bulb which is quite adequate for routine work
(including optical crystallography) and the little 7.5W night
bulb placed in a small receptacle slipped between the feet of
a horseshoe mount microscope and which is much used in clinical and teaching laboratories.

Low voltage bulbs, for example 6V bulbs, are extensively used in built-in illumination systems and for epi-illumination. It is advisable to use the type of bulbs recommended
by the manufacturer of the microscope, in particular because
the mecanical tolerances of a built-in device are rather narrow. Spare bulbs should obviously be kept at hand. In case of
emergency, it is sometimes possible to find usable car lamps
(e.g., stop signal lamps); but the glass envelope of these
bulbs is very irregular and the filaments are small. They are
therefore unsuitable for critical or Köhlerian illumination
procedures[¶] and a strongly diffusing screen must be interposed across the light beam. The glass of the bulb itself may
be made diffusing by rubbing it gently with a fine grade em-

¶   As a matter of fact, these bulbs are perfect for making
the experiment described in 4.1.

ery paper. Note that for field work low voltage bulbs can be
operated from a car battery: a rheostat should be used to
control the intensity of the current (which should be monitored with an ammeter).

Car headlamps have been much used for these procedures
which, like histo- and cyto-spectrophotometry, require a
powerful and steady light source. But there is no need now
to power these light sources by means of rechargeable batte-
ries as it has been done for a long time: this is a bulky
arrangement, and the presence of corrosive and explosive vapors in the laboratory is best avoided. The use of an electronic power supply whose voltage output is regulated by Ze-
ner diodes is much preferable. It should be kept in mind
that the light output of any lamp requires some aging to get
stabilized; a conditioning time of 50 hours is not excessive
for tungsten filament lamps.

Incandescent light sources of the flash type are of
special interest to the microscopist dedicated to long time
studies of live cells: their light output is adequate for
work with most photographic emulsions and the bulb is fired
in so short a time that the effect of deleterious wavelengths
on the biological elements of the preparation is negligible.
But as these lamps can be fired only once, their use is not

practical for automatic operations where, as in time-lapse photomicrography, one wishes to take at regular intervals a picture of the same preparation. In that case one generally employs an electronic flash tube: in this device a gas is excited for a time which may be as short as a few microseconds by a high voltage pulse (for a schematic diagram of the firing circuit see Fig. 4.3.1.). The tube can be repetitively fired at intervals of a few seconds. The color temperature of a flash tube depends on the nature of the gas which is excited: xenon has become popular because the color temperature of a xenon arc is about 5500°K, that is to say that this source is approximately balanced for daylight type emulsions. The main disadvantage of electronic flash tubes resides in their requirement for high voltages: typical values run from 2 to 9 kV; one should therefore be extremely cautious when using them in the not necessarily dry atmosphere of a laboratory.

With the exception of the mercury vapor lamp which is extensively used in UV fluorescence studies[¶], lamps emitting a discontinuous spectrum are seldom used in microscopy, although sodium lamps have been employed in optical crystallo-

---

¶   For a discussion of these light sources, see 4.6.1. and 4.6.3.

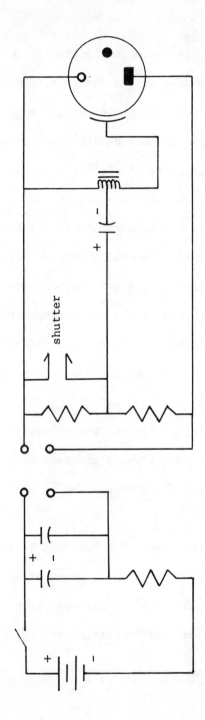

shutter

Fig. 4.3.1. Diagram of an electronic flash circuit. On the left, the power supply: two electrolytic capacitors are charged by a high voltage source. On the right, the flash tube and its firing circuit: when the shutter is closed, the electrolytic capacitor (charged through the voltage divider) is discharged and a pulse is sent to the trigger transformer which fires the flash tube.

graphy. They are rather bulky and evolve much heat; it is therefore advisable for routine work to set up the source and an auxiliary condenser on a small optical bench and to cool the housing of the lamp with a circulation of cold water.

4.3.2.   Filters: generalities.

A filter is essentially a device which excludes from a beam of light rays with certain characteristics: in microscopy the two main types of filters used are those which allow one to isolate light with a certain type of polarization (e.g., linear dichroic filters) and light with certain spectral characteristics (e.g., absorption filters). Note that in this classification one does not take into account the nature of the physical process involved: a linear dichroic filter essentially performs an absorption operation while there exist chromatic filters which rely on interference phenomena and not on absorption mecanisms.

As the output of the light sources suitable for microscopical studies is severely limited, and especially when it comes to built-in illumination systems, the transmission factor of a filter is a parameter often as important as its efficiency. In many instances one has to accept a compromise between the intensity of the light and its purity (e.g., spectral), and this probably explains why interference filters

are much less used than absorption filters. But note that for
the implementation of many a technique of microspectrometry
(e.g., in cytochemistry) a set of interference filters would
constitute an entirely adequate - and less cumbersome - substitute for a monochromator.

Few microscopes do not include at least one absorption
filter as there is almost always a need to remove some IR
rays[¶]. Furthermore chromatic filters are used either to isolate a relatively narrow spectral band (e.g., yellow band for
the measurement of refractive indices by an immersion method,
green band for phase contrast microscopy) or to isolate a
relatively broad band in order to modify the contrast of certain elements of a microscopical object. The latter technique
is valuable in both photomicrography and visual observation:
maximum contrast is obtained with a filter whose color is
complementary of that of the object (for example, a green or
blue object will appear more contrasted in red light) and
conversely details are rendered more apparent by using a filter whose color is similar to that of the object.

[¶]   UV rays of short wavelength are absorbed by the optical
glasses of the auxiliary and of the main condensers. For a
very efficient UV filter, see 4.6.2.

4.3.3.   Laboratory-made filters.

There is a wide variety of commercially available absorption filters, but the fabrication in the laboratory of
special filters is often necessary. In many instances the
microscopist will be at liberty to decide whether to use solid or liquid filters.

Solid absorption filters of the gelatin type are more
permanent and easier to handle than liquid filters; but their
fabrication may be time-consuming, as it is only through experience that one learns to predict the behavior of melted
gelatin. The simplest way to make gelatin filters is to use
photographic plates: the silver halide is removed by treatment in the dark with an aqueous solution (e.g., 25% w/v) of
sodium hypochlorite[¶]. This is followed by a thorough rinse
with tap water and at least one final rinse with distilled
water. The plate is then soaked in an alcoholic solution of
a dye (e.g., aurantia), rinsed and dried. Such a plate may be
used without further ado, but it is easily scratched and for
many applications it is advisable to mount it under a glass
plate, using a liquid medium, and to seal it as it is described below.

¶   Photographic hypo.

A gelatin filter made from a photographic plate is thin; if a thicker (and therefore loaded with more dye) filter is required, a gelatin plate will have to be cast[¶]. As most dyes have slight antiseptic properties, there is generally no problem of conservation. If difficulties are expected in this regard, the addition of cresol (*ca*. 0.25%) could be considered.

Gelatine filters may be sealed with a number of materials: glycerine jelly, household cement, nail polish,... Paraffin wax is not recommended for this purpose; however a mixture of paraffin wax and beeswax, which gives more elastic films,

---

[¶]   A possible procedure is the following one, adapted from the classical method for casting micro-electrophoresis plates. After soaking, the gelatin is dissolved in distilled water (5-10% w/v) in a vacuum filtration flask, over a hot plate set to "low", equipped with a magnetic stirrer. The latter is adjusted so that it revolves very slowly: in this way one can keep to a minimum the number of air bubbles which get entrapped in the medium. When the gelatin has been well dispersed, the dye selected is added in the form of a concentrated solution in water. Once the medium appears to be homogeneous, vacuum is slowly applied to remove the air bubbles. The gelatin is then poured on dry clean plates of glass set on a leveled aluminium plate kept warm with the help of an electric heater located underneath. While the gelatin is still warm, one may directly apply a glass plate on top and let the sandwich cool slowly; this manipulation is difficult with large size plates as obviously one must eliminate (for example with a very fine needle) any air bubble which has been trapped in the process. It is often easier to let the plates cool slowly and once they have reached room temperature to mount them with a liquid medium such as glycerine.

could prove useful.

Liquid filters are at times unstable .but they present the great advantages of cheapness, simplicity and above all versatility: in a few minutes the absorption characteristics can be adjusted (within limits!) to the required values by changing the concentration of some ingredients. However the problem of finding an adequate container is far from trivial when one uses a microscope with a built-in illumination system. Petri dishes, either of glass or of clear plastic, may be mounted on a small cradle made for example with thick wires or aluminium strips; oxidation processes can be somewhat retarded by sealing the cover with an inert compound such as silicone grease. A good quality (optical grade) Carrel flask would certainly prove much superior to - and more manageable than - a Petri dish. When an external light source is used, a rectangular cell (commercially available in many sizes or easily manufactured in the laboratory) is most often employed. It is however worthwhile to recall a method much used by early microscopists: a round flask filled with an acidified solution of copper sulfate was used as an auxiliary condenser. With a 250 or 500 ml round flask and a 200W projection lamp, an improvised condenser suitable for darkfield microscopy and photomicrography can be rigged in a few minutes.

4.3.4.   Light intensity control.

Two kinds of considerations govern the selection and the mode of control of the intensity of the light which reaches the preparation:

1.   the sensitivity of the preparation to certain wave-
     lengths;

2.   the photometric requirements of the detector used.

The photosensitivity of many live biological preparations is well known and in particular UV and IR rays must be care-fully filtered out for long time studies. This is specially important when the preparation must be repetitively exposed to high intensity illumination, as it may be in photomicro-graphy[¶]; in that instance the best solution is to be found in the use of an electronic flash tube. But even for short term studies, the light level may prove to be important, as for example many dyes exert a definite photodynamical action. And inert systems too may undergo photochemical reactions...

The intensity of the illumination can be controlled in a number of ways:

1.   when the light source is a low voltage bulb, the

[¶]   For a review of time-lapse photomicrography, see Paul's book.

voltage applied to the filament may be varied by
means of a multi-tapped transformer: in this way, as
many as half a dozen reproducible[¶] settings of the
light intensity are easily available. But note that
this procedure changes the color temperature of the
source;

2. by the interposition of neutral density filters.
    Some are commercially available and prove ideal when
    they are really achromatic over the entire visible
    spectrum. They can also be constructed in the labo-
    ratory by mounting between two glass plates strips
    of black and white film which have been uniformly
    exposed;

3. by the use of variable density filters. They are
    easily constructed by superposing two linear dichroic
    filters and varying the angle between the directions
    of vibration of the electric vector[§] they transmit:
    the intensity of the transmitted beam is a linear

¶   This presupposes that the voltage applied to the primary
winding is reasonably constant. When this is not the case (in
certain countried variations of the voltage of the main as
high as 40% are not uncommon), some regulation of the input
voltage should be provided for, e.g., with an autotransformer
whose output is monitored with an a.c. voltmeter, or by electronic means.
§   For the terminology used, see 6.1.1.

function of the square of the cosine of that angle.
These devices are very convenient by reason of their
versatility but they present two defects: i) most
dichroic filters are easily destroyed by heat and
therefore IR rays should be carefully eliminated,
for example by means of a plate of heat-absorbing
glass, and, ii) by construction these filters are
not achromatic. In consequence there is at least one
region (the region of the residual hue) in which
they are unsuitable for critical observations.

4.4.   ILLUMINATION METHODS IN BRIGHTFIELD MICROSCOPY.

4.4.1.   Critical illumination.

For a long time this has been a somewhat emotional topic
in microscopy[¶]; partly for semantic reasons: how many bitter
feuds about the exact meaning to be attributed to the epithet
"critical"? Anyway, it is of interest to examine briefly how
and why this method which essentially consists in imaging,
with the condenser, the light source in the plane of the
preparation, came to occupy such a high rank amongst microscopical procedures.

On the theoretical side, it has often been felt that it

[¶]   A role now devolved to Köhler's method of illumination.

comes close to making self-luminous the object under observation and therefore to fulfill one of the conditions of Abbe's famous theory of the microscope; this is not very con-
vincing as the same could be said for example of Köhler's
method of observation. But, experimentally, it was easy to
obtain and pleasant to use with the flame sources (e.g.,
paraffin oil lamps) used at the time.

One can discern a definite evolution in the implementation of critical illumination conditions and this could well
explain, at least partly, some of the lengthy controversies
on the merits and demerits of this procedure. Traditionally
condensers were designed so that a light source located at
approximately 25 cm was imaged 1-2 mm above the stage; final
focusing adjustments were carried out by racking the condenser up and down and/or varying the distance of the extended light source. In this way, one could at one and the
same time obtain a satisfactory illumination of the preparation and control to some extent the size of the illuminated
area as the diameter of the latter is proportional to the
diameter[¶] of the source and inversely proportional to the

[¶]   For purposes of simplification, we consider here a circular source. The extension to the general case (source of
any shape) is immediate.

distance from the source to the condenser focal plane. An
auxiliary condenser of the bull's eye type was sometimes used
to produce a more convergent beam; this presented the double
advantage of restricting the size of the light beam to the
diameter of the mirror and therefore of eliminating a number
of unwanted reflections which might have resulted in glare,
and of increasing substantially the brilliance of the image
of the source.

To be imaged satisfactorily in the plane of the preparation by means of the main condenser only, the filament of
an electric bulb must have a flat ribbon of sufficient size.
This is a rather strict requirement, and especially when
using low power objectives which have a relatively large
field. The use of an auxiliary condenser greatly lessens
this problem, because this accessory can furnish an enlarged
real image of the filament. This provides a means of regulating the size of the illuminated field. But it must be emphasized that the often depicted set up in which the auxiliary condenser furnishes a beam of parallel light is most
inconvenient as [see Fig. 4.4.1.(a)] the size of the illuminated field then solely depends on the size of the filament,
the focal length of the auxiliary condenser and the focal
length of the main condenser, that is to say solely depends

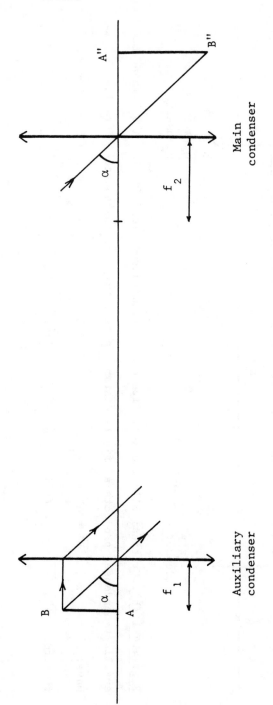

Fig. 4.4.1.(a). Critical illumination: size of the illuminated field (radius A''B'') when the auxiliary condenser furnishes a beam of parallel light; radius of the source: AB. We have:

$\tan \alpha = AB/f_1 = A''B''/f_2$ and therefore $A''B'' = AB \times (f_2/f_1)$.

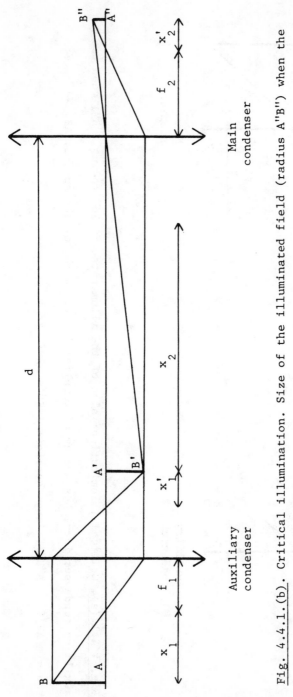

<u>Fig. 4.4.1.(b)</u>. Critical illumination. Size of the illuminated field (radius A"B") when the auxiliary condenser projects at finite distance a real image of the source (of radius AB). We have: $x'_1 = f_1^2/x_1$, $\overline{A'B'} = \overline{AB} \times (x'_1/x_1)$, $x'_2 = f_2^2/x_2$, $\overline{A"B"} = \overline{A'B'} \times (x'_2/x_2)$ and $d = f_1 + x'_1 + x_2 + f_2$;

therefore: $\overline{A"B"}/\overline{AB} = \{f_1 f_2 / [(d - f_1 - f_2)x_1 - f_1^2]\}^2$.

on three parameters which it is bothersome to vary. On the
other hand, if the auxiliary condenser is adjusted to project a real image at finite distance, the size of the illuminated field is also a function of the distance of the filament to the auxiliary condenser and of the separation distance of the two condensers [see Fig. 4.4.1.(b)]: these are
two geometrical parameters which are easily varied.

4.4.2.   Köhler's method of illumination.

This method was originally designed to achieve an easier control, in UV microscopy, of the illumination than it
was feasible with the critical illumination method. To qualify as Köhlerian, an illumination system must satisfy two
conditions:

1. a field diaphragm (in practice an iris diaphragm)
    is imaged by the main condenser in the plane of the
    preparation. This feature allows one to regulate
    the size of the illuminated area;

2. the light source is imaged, by means of an auxiliary
    condenser, in the aperture plane (in practice the
    iris diaphragm plane) of the main condenser.

It may not prove useless to emphasize here that the
second condition is as important as the first one. There is
at times a tendency to equate to a Köhlerian system any il-

lumination system in which a field diaphragm is imaged in the
plane of the preparation (and to carry out the adjustment of
the microscope just that far!). It should be noted that there
is here much more than the mere matter of an arbitrary definition because, as it will be shown later in this section, a
critical phenomenon occurs when the light source is imaged
into the aperture plane of the condenser.

It is often stated that the superiority of the Köhlerian method over the critical illumination method stems from
its unique capability of regulating the size of the illuminated field; but that is not quite correct because, as it is
shown above, this adjustment can be achieved with a critical
illumination system too and therefore, from this particular
standpoint, the selection of a method largely is a matter of
personal preference. The possibility of controlling the size
of the illuminated area through the setting of one comfort-
ably located iris diaphragm appeals to many microscopists
and as precise manipulation of the area of the lighted field
has much to do with the control of glare, the use of Köhler's
method of illumination is usually advocated especially for
photomicrography and phase contrast microscopy. This is hardly a compelling reason for considering it as "the method".

In final analysis, both the most serious weakness and

the greatest advantage of Köhler's method of illumination are
to be found in the demands made upon the light source and e-
ventually its auxiliary condenser: small variations in
brightness of the different points of the light source are
much smoothed over, as a bundle of rays emerging from any
source point exits from the main condenser as a bundle of
parallel rays (see Fig. 4.4.2.), but if the image of the
light source does not fill completely and uniformly the aperture of the condenser, the system is inevitably thrown out
of order: unsymmetrical images and spurious diffraction patterns are commonly observed[¶].

¶   See 4.1.

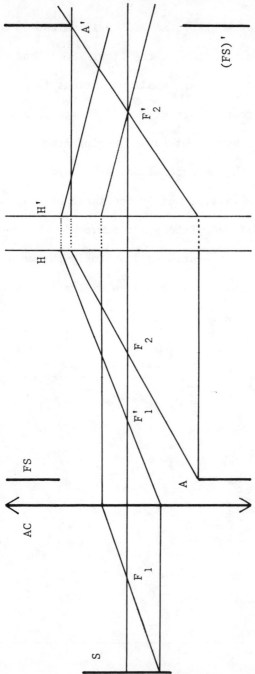

Fig. 4.4.2. Köhler's method of illumination. S: source; AC: auxiliary condenser with its foci F_1 and F_1'; the main condenser is represented by means of its principal planes H and H' and its foci F_2 and F_2'. The image (FS)' of the field stop FS is easily constructed (A and A' are conjugate points. The image of a point of the source is then constructed: it will be noted that all the rays which emerge from one point of the source emerge from the main condenser as a bundle of parallel rays.

4.4.3.   Pseudo-critical illumination.

For a theoretical reason which will shortly appear, we group under this denomination all those illumination procedures which are neither critical nor Köhlerian. Let us consider in the first place what may be called a quasi-critical illumination system, that is to say a system such that the source is nearly imaged in the plane of the preparation. It is easy to see how such a situation can develop: the use as light source of a too small filament is most inconvenient, few microscopists care to observe at the same time the preparation and the image of a coiled filament. Thus it is customary either to unfocus slightly the main condenser or to move slightly backwards the light source, and to interpose a lightly diffusing screen across the illuminating beam.

A second group of pseudo-critical illumination systems includes all the configurations characterized by the presence of a strongly diffusing screen across the illuminating beam.

In all these cases the aperture of the main condenser is filled with diffuse, incoherent light. We take this property as being characteristic of pseudo-critical illumination. Note also that illumination by means of the sole concave mirror often falls in this category.

Many systems labelled critical or Köhlerian are, as a matter of fact, pseudo-critical; this distinction is of importance when it comes to compare the performance of various illumination configurations, in particular with respect to the illumination intensity of the object field.

4.4.4.   Comparison of the different methods of illumination.

All what is really needed here is a set of two propositions established in Born' and Wolf's textbook, namely:

1. the complex degree of coherence[¶] of the light reaching the preparation is the same whether critical or Köhlerian illuminations are used;

2. the complex degree of coherence of the light reaching the preparation in the case of critical illumination is the same as that due to an incoherent source filling the aperture of the main condenser.

The resolving power of an objective is controlled by

¶   Let us consider the wave-field V produced by an extended polychromatic source and two points P_1 and P_2 in this field. The mutual coherence function is the cross-correlation function of $V_1(t)$ and $V_2(t)$, where t is the time. The complex degree of coherence is the normalized (in the sense of the theory of stationary stochastic processes) value of the mutual coherence function. The degree of coherence is the modulus of the complex degree of coherence.

In the two propositions 1 and 2, the expression "complex degree of coherence" may be replaced by the expression "degree of coherence", but they become weakened.

the degree of coherence of the light reaching the preparation
and it follows from proposition 1 that, in this respect, Köhlerian and critical illuminations are strictly equivalent.
In the terminology we propose[¶], proposition 2 means that critical and pseudo-critical illumination systems are completely
equivalent. In other words, from a purely theoretical point
of view, all illumination systems are equivalent.

Experimentally, this is obviously a very different matter as here considerations of the possibility of eliminating
unwanted light rays and the ease of control of the various
parameters involved take a paramount importance.

4.4.5.   Adjustment of the aperture diaphragm of the main
condenser.

This aperture diaphragm controls at one and the same
time two parameters: the convergence of the beam of light
which strikes the preparation and the brightness of the image. But as the latter is also function of the light intensity at the entrance of the main condenser, it is possible,
for a given opening of the iris diaphragm, to adjust the object field brightness through the use of the techniques discussed in 4.3.4. Thus, the condenser iris diaphragm can be

---

[¶]   See 4.4.3.

used exclusively as an aperture diaphragm. In this way, and in this way only, it is possible to adjust optimally micro-scope illumination.

There have been many controversies concerning the best setting of the condenser aperture diaphragm in relation to the numerical aperture of the objective. Roughly speaking, two schools have been arguing the merits of a setting as low as 0.2-0.3 times the NA of the objective *versus* values in the range 0.7-0.8. One may suspect that they were not referring to the same kind of preparation: a low setting increases the detectability of quasi-phase objects while a higher setting is well adapted to the study of amplitude objects (e.g., stained histological sections). Experimentally there are no great difficulties in controlling glare with settings in the range 0.7-0.8 and therefore the latter have been more or less adopted as general purpose settings[¶]. But it must be emphasized that there is no decisive reason to consider these values as necessary upper limits, no more than there exists a physical basis for the feeling that the value of the numerical aperture of the objective constitutes a natu-

---

¶   But the fact that such settings usually furnish images of good quality is no justification for systematically fail-ing to immerse the condenser when using objectives whose numerical aperture is equal to or smaller than 1!

ral upper limit for the numerical aperture of the condenser.
As a matter of fact the results of some theoretical studies
seem to indicate that this implicit assumption is unjustified.
For example, the resolution (calculated using Lord Rayleigh's
criterion) of two pinholes of equal brightness is maximum
when the numerical aperture of the main condenser is approximately 1.5 times that of the objective[¶]. It then appears
that in critical cases it may prove advantageous to explore
a wide range of values for the numerical aperture of the condenser rather than to accept the value of the numerical aperture of the objective as the frontier of a forbidden region. From a practical point of view, it must be recognized
that glare then becomes a very serious problem, but it can
be controlled to a certain extent by restricting the illumination intensity to as low a level as possible, and also by
the judicious use of polarized light.

4.4.6.   Built-in illumination systems.

They are very convenient because of their compactness,
but they are much less versatile than external independent
light sources. In addition their light output is severely
limited because of the intrinsically poor ventilation of the

¶   Hopkins and Barham, quoted by Born and Wolf.

system; in practice, 30W bulbs should not be run at their
maximum rating for any appreciable length of time and for
routine work (continuous duty), 10-15W are to be preferred
in the case of phase contrast microscopy while lower wattage
bulbs (for example 6W) are sufficient for brightfield microscopy. On the other hand, higher wattage electronic flash
tubes can be used if the repetition rate is kept low enough
to ensure an adequate dissipation of the heat generated.

Built-in illumination systems are generally constructed
around a Köhlerian design. There obviously exist many variations in the mecanical construction of these devices but the
system diagrammatically shown on Fig. 4.4.6. may be taken as
representative of the most sophisticated systems available:

1. the field diaphragm is fixed and the sharpness of
    its image in the plane of the preparation is adjust-
    ed by racking the main condenser up and down. With
    standardized preparations (slides of equal thickness),
    the position of this condenser will have to be just
    slightly modified to correct for variations in the
    height of the preparation above the slide. For pre-
    liminary adjustments, a screen made by mounting in
    air (under a coverslip) a fragment of tracing paper
    proves convenient; final adjustments are made while

observing directly the image of the field diaphragm
in the microscope;

2. there is no centering device for the light bulb.
    This is made possible by the use of bulbs built to
    very close mecanical tolerances. These bulbs are
    often said to be prefocused; in most cases however
    a small adjustment of the distance between the bulb
    and the first lens of the auxiliary condenser will
    be required;

3. the mirror M, inclined at 45° with respect to the
    axis AA', has two degrees of freedom, one of trans-
    lation along AA' and one of rotation around the same
    axis. When the mirror is properly positioned, the
    intersection O of the axis AA' and of the axis BB'
    (= optical axis of the microscope) is in the plane
    of the mirror, and the normal Ob to the latter plane
    is in the plane defined by AA' and BB': then the im-
    age of the field diaphragm FS (assumed centered with
    respect to AA') projected by the main condenser
    (previously centered) in the plane of the preparation
    will be centered with respect to BB', that is to say
    will be centered in the object field.

The latter property may be put to use to adjust the sys-

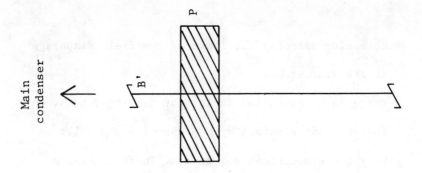

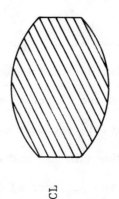

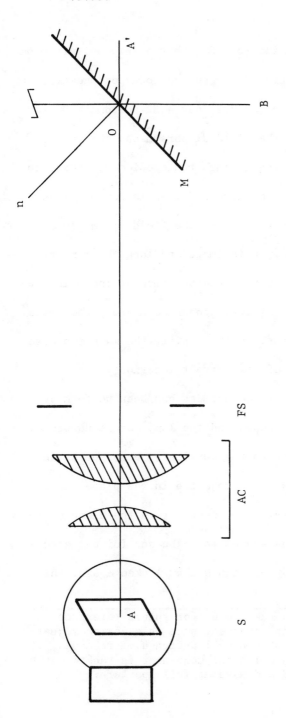

Fig. 4.4.6. Diagram (not on scale) of a built-in illumination system. S: light source; AC: auxiliary condenser; FS: iris diaphragm functioning as field stop; M: mirror; P: exit window; CL: converging lens which can be substituted for the exit window P; n: normal to the plane of the mirror M.

tem for Köhlerian illumination. Switch the lamp on, place on
the stage the aforementioned screen and focus on the tracing
paper with a low power objective, say a 3X achromat. Unlock
the mirror, close down the field diaphragm and the aperture
diaphragm. While observing through the eyepieces, impart to
the mirror movements of rotation and translation until an
image - most likely not sharp - of the field diaphragm is
perceived. Center crudely this image and lock the mirror.
Sharpen the image by adjusting the position of the main condenser; then unlock the mirror, center accurately the image
and lock again the mirror. At this point, the main condenser
is in the neighborhood of its correct location[¶].

Once this preliminary adjustment has been performed,
the field diaphragm is opened and the lamp socket unlocked;
the distance of the lamp to the auxiliary condenser is then
adjusted so that an image is projected in the plane of the
iris diaphragm of the main condenser (or in the plane of the
annular diaphragm if the system is to be set for phase contrast illumination) and the lamp socket locked again: this

---

[¶]   It is advisable to take note, for future reference, of
the position of the main condenser once this coarse adjustment has been carried out. In well designed microscopes, it
is always a position in which the condenser is racked way up:
thus it can be immersed and used at full aperture.

image is best[¶] inspected with the help of a small plane mirror positioned near the exit window at a small angle with the
horizontal plane. The aperture diaphragm may be completely
closed in order to function as a screen, but it is often
shiny and it may prove difficult to obtain a satisfactory
image; the same occurs with the annular diaphragms which are
often sandwiched between glass plates for mecanical protection. In such a case a bit of tracing paper will constitute
an efficient screen. At the end of this step, Köhler's conditions are fairly well approximated: they can be definitely
established by readjusting slightly the position of the main
condenser while observing in the microscope the image of the
field diaphragm and in the back focal plane of the objective
the image of the light source.

The presence of a mirror with two degrees of freedom is
most advantageous as, in particular, this feature enables
the observer to switch easily from purely symmetrical to ob-

[¶]   Obviously this image can also be observed by looking at
the back focal plane of the objective (which is conjugate of
the aperture plane of the main condenser; see 6.6.2.). However this procedure should best be kept for carrying out
minor adjustments - if need be - because the Bertrand oculars
commonly used for this purpose do not give very sharp images.
The microscopist is strongly urged not to take for granted
the quality of the illumination system and, from time to
time, to inspect directly the image of the filament as described here.

lique illumination. But the precision mecanisms required for the smooth control of the position of this mirror are expensive and therefore simpler apparatus are built around another design, physically equivalent and adequate so long as the observer accepts the limitations associated with the use of a strictly symmetrical illumination method: the mirror M is now fixed and the lamp is set in a centerable mount. The adjustment procedures are carried out as described above, substituting a lateral displacement of the light bulb in its mount for a rotation of the mirror.

Solid filters can be positioned between the auxiliary condenser and the field diaphragm; although this location is excellent from an optical point of view, it presents certain disadvantages:

1. the filters have to be cut to size and mounted in special mecanical holders;

2. this is a poorly ventilated area and gelatin filters would be rapidly damaged;

3. it is inconvenient to have to remove the auxiliary condenser assembly in order to be able to change one filter.

It follows that this location should be used only when one plans to use continuously the same filter.

The light beam is most easily accessed in the region between the exit window and the main condenser. Filters and polarizing devices can be placed directly on the exit window or supported by cradles.

Ideally, ribbon filament light bulbs should be used, but they are not available in all sizes and manufacturers leans towards the use of tightly coiled filaments. Under these circumstances it is usually necessary to pseudo-criticalize the illumination. Note that stray light will be somewhat decreased if the diffusing screen is positioned close to the condenser aperture; when the condenser does not include a filter holder[¶], this screen may be supported by a small metal cylinder bearing on the frame of the exit window. But one should be suspicious of this new brand of auxiliary condensers in which the first air-glass interface has been ground so that it functions as a diffusing screen: a much better adjustment of the illuminating beam will be achieved if one proceeds in two successive steps, first setting up conditions of Köhlerian illumination and later, if need be, pseudo-criticalizing the latter.

It will often be found that built-in illumination sys-

[¶]   This is in particular the case of the phase contrast condensers fitted aith an annuli-bearing turret.

tems of a Köhlerian configuration furnish a lighted area
smaller than the object field of some low power objectives.
To fill evenly the latter the system is set in a critical
(or near critical) mode: the flat exit window is replaced by
a converging lens (see Fig. 4.4.6.).

4.4.7.   Epi-illumination.

The preceding discussion of critical, Köhlerian and
pseudo-critical illumination methods is equally valid for
trans- and epi-configurations, but certain experimental details particular to the latter must be pointed out.

With vertical illuminators, critical illumination conditions are easy to realize as one can directly observe the
image of the light source in the plane of the preparation.
But true Köhlerian illumination is more difficult to attain.
First, the field diaphragm must be imaged by the objective
in the plane of the preparation: this presents little difficulty because, exactly as in the case of critical illumination, the real image so obtained can be observed directly
through the microscope. But in addition, the light source
must be imaged, by means of an auxiliary condenser, in the
back focal plane of the objective (focal plane which plays
here the same role as the first focal plane of the main condenser in a trans-illuminated system): in practice, for rea-

sons already mentioned[¶], this requires the temporary insertion of a screen, for instance a disk of tracing paper, on
the objective diaphragm.

With low power objectives, lateral illumination by means
of mirrors, either plane or concave, is often efficient.
When, as it is usually the case, the illumination is unsymmetrical, some shadows are produced, which results in a 3-D
effect. The mirrors can be clamped to a regular laboratory
stand or affixed to the microscope stage by means of tiny C-
clamps or with a lump of plasticine. A useful accessory is
a laryngoscope[§]: this instrument can be used hand-held to explore rapidly the effect on the image of many different directions of illumination, and set in a laboratory stand with
the mirror locked by means of a bit of masking tape for prolonged observation and for photomicrography.

Small light sources suitable for epi-illumination are
easily improvised. In particular, in many large research microscopes with a built-in illumination system, the lamp and
the auxiliary condenser are mounted as a subassembly in a
sleeve which slides in the base of the instrument and is
locked in place by means of a screw with a protruding knurled

---

¶   See footnote p.179.
§   Available from most medical supplies houses.

head. This subsystem constitutes a source very efficient for
the purpose at hand. Note however that when the auxiliary
condenser assembly is removed for any appreciable length of
time, it should be replaced by a metal or plastic plug in
order to avoid the deposition of dust on the field diaphragm
and the 45° mirror.

Specially with low power objectives, light guides[¶] may
prove useful: rigid guides can be made of slightly tapered
polymetacrylate cylinders[§], curved in a steam bath if need
be, or of commercially available glass fibers bent in the
flame of a Bunsen or Mecker burner; flexible glass- and plas-
tic-fiber light guides are also commercially available.

4.5.   ULTRAMICROSCOPY AND DARKFIELD MICROSCOPY.

4.5.1.   The place of darkfield microscopy.

A critical survey of the literature reveals that darkfield procedures have been comparatively little used though
their intrinsic merits were early recognized. It seems that
in general microscopists do little more than take cognizance
of their existence and at most pay them lip service. For in-

[¶]   Note that their use is not restricted to epi-illumination. For example, the trans-illumination with a rigid light
guide of thin biological structures (interdigital skin of
the frog, intestine strips,...) is often convenient for observation and dissection.
[§]   Stray light effects can be eliminated by wrapping such
guides with black plastic tape.

stance, in most manuals of clinical bacteriology and histo-
pathology, the existence of darkfield microscopy is acknow-
ledged in connection with the detection of *Treponema palli-
dum*; and in textbooks of physical chemistry one will usually
find mentioned the classical studies of Zsigmondy and Siedentopf; but precious little else. This lack of interest appears
steadily maintained in spite of the very favorable comments
of contemporary writers: Paul pointed out that this technique should prove useful in tissue culture studies, Chamot
and Mason have repeatedly insisted on the value of systematic
examinations of microchemical samples by darkfield microscopy, etc... The use of darkfield microscopy is well established in certain technological domains such as metallurgy,
but in biological research and in biomedical investigations
little interest has been shown for this technique during the
last few decades with but two exceptions: as a visualization
technique for ashes in Policard's micro-incineration procedure and, to a slight extent, in fluorescence microscopy.
This disaffection has been attributed to the appearance on
the scene of phase contrast microscopy, but not too long ago
Barer was deploring the loss of interest in phase contrast
procedures!

This state of affairs is most regretable. The difficulties of successful darkfield microscopy, although real,

have been overemphasized. Once it has been granted that from

a practical point of view there is some overlapping between

darkfield and phase contrast procedures due to the import-

ance in both cases of the difference between the refractive

index of the particle and that of its surrounding medium, it

must be pointed out that these two techniques deal with some-

what different properties of matter and that each proves most

useful in rather different circumstances: the halo well known

to phase contrast microscopists and generally considered ob-

noxious (but see Chapter 1) disappears or at least is consid-

erably reduced when the refractive index of the particle is

close to that of the surrounding medium, while in the same

conditions in darkfield microscopy the particle "disappears".

Let us add also, for completeness, that in the submicroscopic

range the diameter of resolvable particles is smaller with

darkfield procedures than with variable phase contrast tech-

niques. Under these circumstances, it appears that a discus-

sion of darkfield illumination systems is called for, and

the more so when it is realized that modern equipment and

preparation procedures which have become routine for the

phase contrast microscopist are quite adequate (with trivial

modifications) for the darkfield study of a variety of samples.

This should not be construed as a plea by the author to consider darkfield procedures as a panacea, but they certainly should be accepted as an integral part of the current arsenal of the microscopist. The problems one faces in microscopical investigation are so complex that to refuse the benefits of another non-destructive technique is tantamount to self-mutilation.

4.5.2.   Ultramicroscopical methods.

Traditionally, following Siedentopf and Zsigmondy whose interest was directed towards the detection of particles invisible with the standard techniques of the time, a distinction is made between ultramicroscopical and darkfield techniques, and there is a strong tendency to regard the latter as the "parents pauvres" of the former. This attitude is hardly defensible anymore as the physical principles involved in the illumination of the sample is the same in both cases; it is mainly in the interpretation of the image in relation to the size of the diffracting units that differences may be found. And as a matter of fact the same device used in the same way can often be labelled an ultramicroscopical condenser or a darkfield condenser according to the nature of the sample with which it is used: this is for instance the case

with the cardioid condenser which can be employed for the
study of fine colloidal suspensions or of biological preparations.

Thanks to the ingenuity of the physicists, the microscopist now has at his disposition a number of devices specifically designed for fulfilling the requirements of darkfield microscopy. In terms of detection limit, the immersion
ultramicroscope of Zsigmondy is probably the finest, but it
is difficult to set up and manipulate, and its use is mainly
restricted to the study of non-corrosive hydrosols and of
dispersions in an oil base. The slit ultramicroscope of Siedentopf and Zsigmondy, although it can be used with a variety
of materials, presents serious disadvantages too: it is not
of an easy manipulation and fluid samples have to be contained within special cells such as the Biltz cells which are
fragile, costly and difficult to clean. In practice, the use
of these two ultramicroscopes is restricted to highly specialized laboratories and therefore their discussion falls
outside the scope of this book. There exists also a very
clever, but apparently forgotten, ultramicroscopical device
which will be mentioned briefly here because it deserves, due
to its simplicity, to be brought again to the attention of
the microscopists. This apparatus, the prism of Cotton and

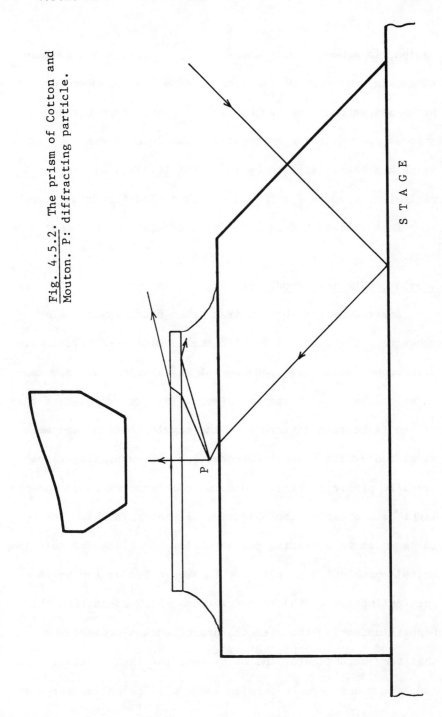

Fig. 4.5.2. The prism of Cotton and Mouton. P: diffracting particle.

Mouton, is schematically shown on Fig. 4.5.2. The illuminating beam undergoes reflections first on the base of the
prism and secondly on the coverslip. Under these circumstances, the microscope objective ideally collects only light
which has been scattered by the particle. This system was o-
riginally designed for the enumeration of diffracting granules in a colloid but is clearly adaptable to many other systems.

4.5.3.   Darkfield condensers.

Apart from the three ultramicroscopical systems aforementioned, all others (to which the name of darkfield condensers properly applies) make use of an illuminating beam centered on the axis of the microscope. The strong point of this
design is to take advantage of the regular optical and mecanical features of the microscope, but not surprisingly some
specific problems arise. The main goal is to exclude from the
light beam entering the objective all the rays which belong
to the beam illuminating the particle. The classical solution
to this problem is to effect this separation on the basis of
the inclination of the rays: if those which constitute the
beam illuminating the particle are of such a convergence
that they cannot enter the objective, the light collected by
the latter and used to form an image will be exclusively part

of the beam diffracted by the particle. This is to say that the phenomenon will be controlled by the numerical aperture of the objective.

It will be noted that various phenomena of total reflection at the coverslip interface(s) [which depend on the refractive indices of the immersion medium, the coverslip and the medium between the latter and the front lens of the objective] contribute to eliminate from the beam which exits from the preparation, rays which make a large angle with the optical axis[¶].

Darkfield condensers are traditionally classified as reflecting or non-reflecting according to the existence or the absence of internal reflections within the condenser itself. This classification was obviously designed in regard to the condensers used with trans-illuminated preparations, but it is easily extended to include all darkfield condensers as darkfield vertical condensers always incorporate some kind of reflecting device. At any rate it must be emphasized that all darkfield condensers, whether or not they are of the reflecting type, include a central stop in order to eliminate part of the entrance beam and to allow to pass

---

¶   For some examples, see Fig. 4.5.2.

through only those rays to which the condenser can impart a convergence greater than the half-angle of acceptance of the objective.

Amongst the reflecting darkfield condensers intended for use with trans-illuminated preparations, the cardioid condenser and the paraboloid condenser, both designed by Siedentopf, are probably the best known and the most popular. The cardioid condenser has proved to be the instrument of choice for the study of delicate structures (for instance in histology); unfortunately it is a complicated device to manufacture as it makes use of reflections on two spherical surfaces and therefore it is rather expensive. The paraboloid condenser has been successfully used in a number of fields; it is made from a paraboloid truncated by a plane perpendicular to the axis of revolution. The preparation is positioned in such a way that the element under observation is exactly at the focus of the paraboloid; as this point is stigmatic for rays parallel to the axis, there is an intense concentration of light, a situation most favorable to the detection of very fine details.

In non-reflecting condensers[¶], one relies mainly on the

[¶]   Which are exclusively used with trans-illuminated preparations.

central stop. This type of device suffers from a bad reputa-
tion because with an Abbe condenser it is not very efficient
at high numerical apertures and the contrast is poor. This
difficulty is attributable to the lack of aplanatism of an
Abbe condenser. In conjunction with an aplanatic condenser,
a central stop performs well and for routine work it proves
quite satisfactory. To this group belong too the stage condensers which have been too hurriedly condemned into oblivion; as their name indicates, they are (darkfield) condensers
intended to be set upon the stage; therefore they can be used
with most microscopes. They include two optical groups and
should be illuminated with parallel light. In the better
models, one group is mounted within a barrel sliding along
an helicoidal ramp: thus the condenser has a variable focal
length[¶]. It follows that the working distance is variable to
some extent and that the apparatus can be used with slides
of any thickness. The advantages derived from the presence
of this feature (sorely lacking in some of the more sophisticated designs) more than compensate their main drawback,
that is to say the risk that, due to a careless manipulation,
they accidently become decentered. The latter occurrence is

[¶]   Such a condenser is sometimes referred to as a "variable focus" condenser.

nearly inevitable if one follows the usual procedure of lock-
ing them into place by means of standard stage clips; a much
better way is to secure them with large size bulldog clips.

It will be noticed that all the darkfield condensers
discussed here essentially make use of an annular cone of
light. This is exactly the type of illumination for which the
modern Zernike phase contrast condensers are designed. It
follows that the phase contrast microscopist has at his fingertips the tools required for routine darkfield illumination.

4.5.4.   Resolution in darkfield microscopy.

The resolution of an ultramicroscopical or darkfield
system depends upon:

1. the quality of the illumination;

2. the intensity of the source. The sun is still an un-
    surpassed source for the critical ultramicroscopy
    of extremely fine colloidal systems, but thanks to
    the steady improvements brought to light sources
    and auxiliary condensers during the last decades,
    darkfield microscopy has become increasingly easier;

3. the difference between the indices of refraction of
    an element and of its surrounding medium. If this
    difference is zero, the element or the particle
    will not be distinguishable from the background;

this is a phenomenon well known in colloidal chemistry where the class of the isocolloids was early created on the basis of this observation. Few difficulties are encountered in chemical microscopy because, as a rule, crystals have higher indices of refraction than their mother liquor, and anyway it is generally easy to find another suitable mounting fluid of a very different refractive index, as the only considerations of importance here are ones of solubility and chemical inertness. In biology, on the other hand, the situation is slightly different: the average refractive index generally is but little higher than that of the surrounding medium[¶]; in most cases it will be easy to secure adequate images. However the difference of indices may prove insufficient for critical studies. In that event one should try to achieve conditions such that the difference between the refractive index of say the cytoplasm and that of the mounting medium is as large as possible and at the same time such that an ap-

---

¶   But there are exceptions. A classical example due to Barer and Ross is that of earthworm amoebocytes whose cytoplasm has a lower index of refraction than blood.

proximately biological environment is maintained for a time long enough to permit meaningful observations. The solution of this problem is to be found in the use of artificial iso-osmotic fluids and not of "natural " media such as serum, tissue culture media, etc... which contain proteins, because the latter substantially increase the refractive index of the solution (at room temperature and for yellow light, $\Delta n/\Delta c = 0.0018$, where n is the refractive index and c the protein concentration expressed in grams per 100 ml).

Let us add that the arguments developed in 3.2.4. about the resolving power of an objective used with brightfield illumination are equally valid when one resorts to darkfield illumination.

4.5.5.   Immersion of darkfield condensers.

The time-honored advice to the microscopist to immerse darkfield condensers is supported by a number of arguments, in particular:

1. it is clearly imperative if one wishes to illuminate the element observed with very oblique rays;

2. this procedure eliminates for the greatest part the diffraction of light by the scratches which may be

present on the underside of the slide; as a result,
there is a lesser loss of light at this interface
and the background tend to be darker.

On the other hand, when darkfield illumination conditions are realized with the help of an annular diaphragm
(such as those used in phase contrast microscopy) whose external radius corresponds to a numerical aperture less than
unity, immersion of the condenser is not mandatory if it is
aplanatic.

4.5.6.   Rheinberg illumination.

A very clever modification and extension of darkfield
illumination is due to Rheinberg, and is generally referred
to as Rheinberg illumination or differential color illumination.

In this system the central opaque stop of a darkfield
condenser is replaced by a colored (color A) and more or less
transparent disk, and the exterior clear ring by a colored
(color B) ring. In the conditions of darkfield illumination,
that is to say when the image of the central disk covers the
whole exit pupil, a particle will appear of color B on a
background of color A. A Rheinberg filter is easily made from
a rigid cellophane sheet and/or water colors and poster colors.

As the central stop is not necessarily opaque, Rheinberg illumination method does not depend entirely on diffraction phenomena for the visualization of a particle, but partially too on absorption phenomena[¶]. The relative importance of absorption and diffraction effects, their balance, can be controlled by varying the degree of opacity of the central disk.

Rheinberg's method can be useful for instance for the observation of micro-crystals, as often facets difficult to distinguish with conventional methods may be made to exhibit dramatic changes of color. Note that many variants are feasible: for example the ring may be made of two different colors to allow one to discriminate between adjacent micro-facets. In the latter instance the use of a rotating stage is very convenient.

4.6.   FLUORESCENCE MICROSCOPY.

4.6.1.   Generalities.

In brightfield fluorescence microscopy with transmitted light, a given UV line is isolated by means of a filter (primary filter or exciter filter) and a beam of adequate convergence is focused, for instance by means of a quartz condenser, on the preparation. In these conditions the light

---

[¶]   See also 4.1.

which is collected by the objective is polychromatic: in addition to the visible component(s) due to the fluorescent
elements of the object under investigation, the exciting UV
line is also present. If the latter is intense, which is
usually the case, special objectives characterized by the
absence of fluorescent components, must be used, especially
for high power work. The exciting line is absorbed by a filter (secondary filter or barrier filter) so that the image
appears bright on a dark background; the secondary filter
must remove absolutely all the UV components from the beam
in order to prevent damages to the eye and fogging of the
photographic emulsion. This filter may be located in a number of places; if the working distance is large it can be
interposed between the preparation and the objective, and in
that case obviously all the restrictions about the presence
of fluorescent components within the objective are lifted.
But most often the secondary filter is placed between the
objective and the eyepiece, within the eyepiece (for instance
supported by the field diaphragm), or cap-mounted.

The procedure just described leads to the formation of
a bright image on a dark background. A very efficient way to
arrive to that result is clearly to use a darkfield condenser.
This technique, which has not found great favor with fluo-

rescence microscopists, presents the added benefit that, once the darkfield conditions have been met, only a (most usually) negligible amount of UV light enters the objective: a fraction of the UV light scattered forwards by the preparation. Therefore regular objectives can be used for all magnifications; in the same way, the secondary filter can be dispensed with during observation and photography. Note however that with an intense UV source, it is safer to use an UV absorbing filter while adjusting the apparatus.

4.6.2.   Safety precautions.

Any UV beam, even of low intensity, must always be treated with utmost respect as retinal damages can be serious (*cf.* the painful snow blindness or niphablepsia, or the "eclipse blindness"). There is little warning (although some fluorescence of the vitreous humor has been reported by a number of investigators) and once pain develops, it is too late and nothing much can be done but to consult a good ophtalmologist. If reasonable care is exercised, no mishap will happen; nevertheless students and technical personal should be repeatedly reminded to keep on the lookout for possible "leaks" of UV light.

Unwanted reflections of the UV beam may be prevented by baffles and protective screens. Extemporaneous baffles, often needed when the preparation is illuminated by reflection, are easily constructed from heavy duty aluminium foil. The absence of leaks may be checked with fluorescent preparations, for instance with slides on which a little vaseline has been smeared. The blue fluorescence of human skin is a reliable index of the presence of UV light too! but the hands should not be used as a substitute for a physical monitor as an erythema could easily develop.

When the fluorescence microscope is not used with a darkfield condenser, an efficient UV filter must be interposed across the beam. Glass filters are commercially available[¶], but many investigators still rely on the excellent liquid filter suggested by Gründsteidl in 1933: an aqueous solution of sodium nitrite. A few millimeters of a 2% solution are sufficient.

4.6.3.   Equipment.

From a technical point of view, the limiting factor in fluorescence microscopy is often the output of the light source which is usually a mercury-vapor lamp. The emission spectrum of this device sonsists of strong lines at 2537, 3129, 3654, 5461 and 5780 $\mathring{A}$ and of a more or less (depending

---

[¶]   And are used as barrier filters in specialized fluorescence microscopes.

on the pressure) continuous background. The interest is main-

ly focused on the UV lines, but in spite of their apparently

high intensity, the efficiency of the system is inherently

low: even with a high-pressure lamp operating at 1.2 atmos-

pheres, one of the most powerful sources utilized in a micro-

scopy laboratory, less than 3% of the input power will be

found in the 3654 Å line[¶]. This explains why high intensity

sources have to be used, especially when one insists upon a

binocular mode of observation; lamps of several hundred watts

are far from uncommon and the cooling of these devices pre-

sents serious problems. Needless to say that these lamps can-

not be used in built-in illumination systems: they are always

in external housings and the exit beam, preferably confined

within a metal tube, is deflected towards the main condenser

by means of a first surface mirror[§]. Attention should be paid

to the mecanical design and to the material of the lamp hous-

ing: models made of metal and covered with external radiating

ribs are particularly efficient.

¶   This line is of special interest because it excites, in
particular, the fluorescence of acridine orange, a water-
soluble dye which is used in a number of procedures, for example the discrimination of DNA and RNA in cytology (Berta-
lanffy's method), the staining of mucin after ferric chloride
mordanting and the vital staining of soil bacteria.
§   Alternatively, one can use a total reflection prism made
of quartz.

The use of well isolated UV lines is quasi-mandatory for semi-quantitative and quantitative studies, and may be advantageous for qualitative observations as a higher specificity is then achievable. As a result there is nowadays a strong tendency to equate fluorescence microscopy with the use of very narrow UV spectral bands. It must be emphasized however that the fluorescence of certain substances is excited by blue light of short wavelengths and that a source with a composite spectral distribution is useful for revealing simultaneously many fluorescent elements in a preparation: such studies are best conducted with a monocular microscope and a low-pressure mercury vapor lamp. The system is adjusted either for epi-illumination or for darkfield transillumination. In the latter instance light losses can be reduced to a large extent by mounting the light bulb (and its firing circuit) in a wooden box upon which the microscope is secured: an aperture is cut in the top of the box so that the light beam can enter the condenser directly.

4.6.4.   Focusing aids.

The fluorescence of uranium glass has long been recommended as an useful aid in focusing condensers and specially darkfield condensers; small blocks and thick plates are commercially available. Vaseline (white petroleum jelly) may also be found convenient for this purpose; its fluorescence color is a function of its thickness: blue for very thin layers and green for thick smears. Obviously, any strongly fluorescent material may be used, for instance in the near UV acridine orange which absorbs at 365 nm and emits at 542 nm [in the green region of the spectrum, see Table 4.3.1.(b)], is easily available. The use of magnesium platinocyanide has been recommended by George Needham.

Even when no focusing problem is expected, it is usually advisable to do as much prefocusing as possible with an auxiliary preparation, because the fluorescence of many substances rapidly fade; quite a few biological compounds[¶] are notorious in this respect. It is therefore a wise precaution to treat an unfamiliar fluorescent system as if it were fragile and not to expose it to an UV beam for a long period of time before starting the observations.

¶   For example, vitamin A.

CHAPTER 5

MICROPROJECTION AND PHOTOMICROGRAPHY

5.1. MICROPROJECTION.

Although any microscopical procedure essentially involves the projection of an image, either on the retina, or on a photographic emulsion, or on the entrance slit of a spectrograph, etc..., the use of the word "microprojection" is traditionally restricted to describe those configurations in which an image is projected on a screen where it will be visually observed. Microprojection techniques are much less used nowadays for demonstration purposes than formerly as it is easier to project transparencies than preparations, but they remain invaluable for example for quality control operations, and for profile measurements in the tool room. Note also that in the determination of discrete constituents (for instance, analysis of mixtures in the food industry), it is less tiring for the operator to perform counts on an image than through an eyepiece, and more rapid and less costly than to work on a photomicrograph.

Any microscope can be converted into a microprojection apparatus by the addition of a high intensity light source and of an efficient IR filtering device such as a thick water cell. The optical system used is similar to those used in photomicrography and, in addition, most often includes mirror(s) or a total reflection prism to deflect the beam towards the screen. The quality of the image varies much with the intrinsic magnification of the microscope, the size of the image and the nature of the screen. When all that is required is a projected image of relatively small size, a translucent screen (for example a sheet of matte cellulose acetate or a plastic Fresnel lens) will prove convenient; with such a device high power objectives will give acceptable images, although one should expect some loss of resolution. With regular ground-screens or cinema-type screens, the degradation of the image is rather severe.

5.2.   EQUIPMENT FOR PHOTOMICROGRAPHY AND TECHNIQUES.

5.2.1.   Selection of a camera and setting of the microscope.

There are few microscopists who have not felt the need to record photographically the image they visually observe. Nowadays it is no great feat to rig up a camera on a microscope and, after a few trials, to obtain a picture which bears more or less resemblance to what was seen through the

the ocular. Needless to say that such a casual attitude is not going to lead to very satisfactory results.

The necessity of making systematic photomicrographs could well be questioned in many instances. There is little doubt that since the advent of the trinocular body equipped with a photographic head (a godsend to the responsible micro-scopist), a fantastic amount of film has gone to waste, which is highly regrettable; what is even more regrettable is that roughly proportional amounts of time and energy have been wasted too. It must be emphasized that the eventual value of a photomicrograph as an objective document should be weighted in each case, because in addition to the usual difficulties of deciding what constitutes a representative field, photo-graphic limitations due for example to the spectral sensiti-vity of the emulsion and the very reduced depth of field are also to be taken into account.

The 35 mm (135 format) films are very popular today for photomicrography, and this for a number of reasons: they are relatively cheap and are available in a variety of emulsion types, they can be handled and processed without a darkroom (a changing bag is sufficient), many 35 mm reflex cameras are easily adapted for this purpose, and negative rolls and trans-parencies are conveniently filed in a small space. However,

because of the small size (24.5 × 36.3 mm) of the negative,
high magnification ratios are necessary in order to obtain
readable prints; this requires that the negatives be of a
very high quality. One can easily conceive of situations in
which the use of other film formats would be advisable: for
instance, if prints of a very large size were required, as
is the case for museum displays. But as the techniques used
in such circumstances are very similar to those employed
with 35 mm films, we shall exclusively concentrate on the
latter. Mention must be made however of the use of Polaroid
films in photomicrography; they have been utilized with success by a number of investigators, but they still present a
few disadvantages, not to mention their price: their routine use in microscopy proves to be an expensive endeavor.

Whatever may be the model of camera adopted, its weight
should be matched to that of the microscope if the apparatus
is used in an upright position. In the first place it is advisable to keep the center of gravity of the whole system as
low as possible; if this does not prove feasible the camera
should be supported by a stand. Although one could use a
laboratory universal stand, it is much more convenient to
employ the stand of a small photographic enlarger: the camera can be directly screwed on the slide arm in place of the

enlarger head. The second, and perhaps more important, consideration is that of the load on the micrometric screw which should be kept small[¶].

In certain locations vibrations can constitute a most serious problem. Often it will prove sufficient to set up the microscope on a wooden board loaded with weights and whose base bears on a mattress of felt or, better, on soft rubber strips (self-adhesive weather strips are an adequate material); alternatively, hemispherical rubber balls, either of foam or cut from tennis balls[§], can be used to support the board and dampen vibrations.

In worse cases one has to resort to the technique used with sensitive balances and galvanometers, and to set the apparatus on a plate of lead alloyed with antimony (for instance the lead alloy used in the printing industry) located on a bed of sand; the letter should have been purified by elutriation in order to remove the fines as a source of abrasive dust is unwelcome near a microscope.

Abrupt movements of the shutter cable may also create vibrations and therefore it has become customary to use a not

---

¶   See 2.1.1.
§   Rubber is easily cut with a sharp knife wetted with cold water.

too short release cable. When there is more than one release
cable, for instance in the case of a trinocular microscope
with a photographic head, it is advisable to get these cables
out of the way, but in no case should they be tightly fixed
to the microscope. Split plastic rings or wide split rings
made of soft rubber are useful to position loosely these cables, as well as related accessories such as flash connecting
cables.

5.2.2.   Adaptation of a single-lens reflex camera.

Most single-lens reflex cameras with a non-integral lens
are suitable for photomicrography; a set of extension rings
is a useful adjunct which permits one to vary the distance
between the eyepiece and the emulsion plane. In this way, a
certain control of the magnification is made possible and,
more important in practice, this often allows one to minimize
various aberrations.

One type of attachment which is commercially available
consists of a metal ring to be permanently fixed to the drawtube (or equivalent); to this ring is attached, by means of
a small hinge, a second ring which is permanently fixed to
the camera extension tube. The position of the two rings is
adjusted so that the camera can be swung above the eyepiece
or beside the microscope body. Practically, the use of this

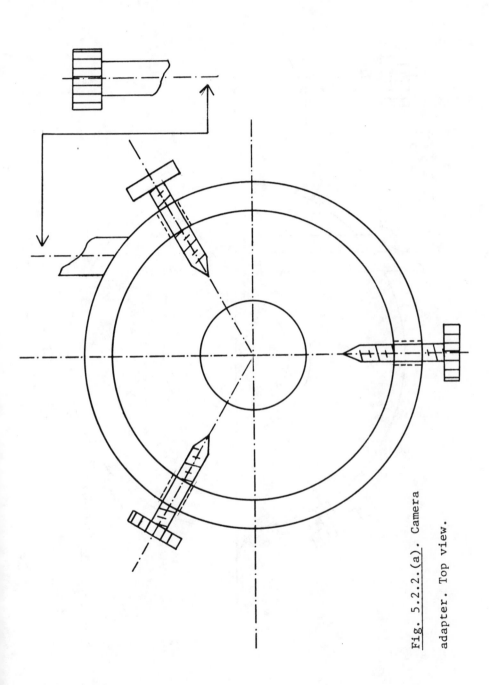

Fig. 5.2.2.(a). Camera
adapter. Top view.

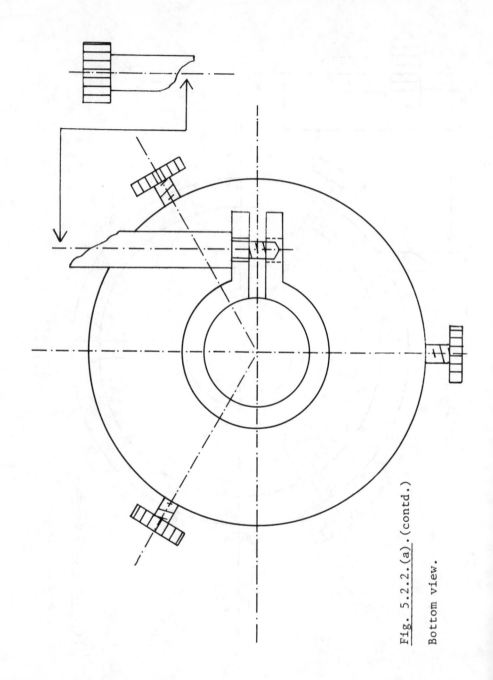

Fig. 5.2.2.(a).(contd.)

Bottom view.

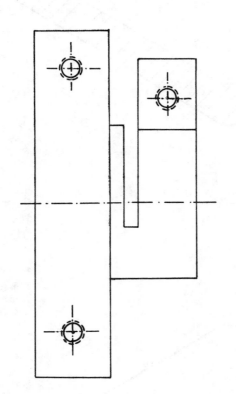

Fig. 5.2.2.(a). (contd.). Side view; the knurled-head screws have been removed.

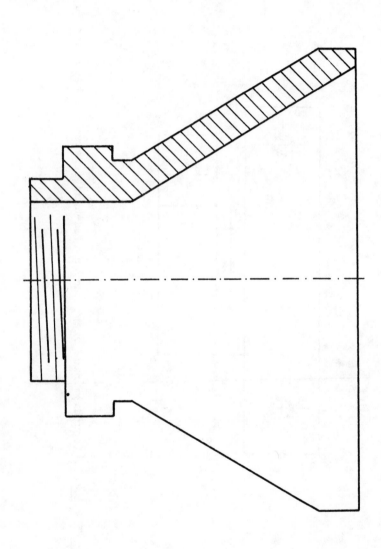

Fig. 5.2.2.(b). End-piece to be screwed to the extremity of a camera extension tube for use with an adapter such as the one depicted on Fig. 5.2.2.(a). Composite side view and cross-section.

system is restricted to very light cameras.

By far, the most convenient way to attach a reflex camera to a microscope is by means of a rapid exchange mount.
Some systems are commercially available; when problems arise
for matching a given camera to a given microscope, the accessory shown in Fig. 5.2.2.(a) can be easily machined from
brass stock in the laboratory machine-shop. An end-piece
such as the one shown on Fig. 5.2.2.(b) is screwed at the
extremity of the camera extension tube: it is held in position by means of three screws (with knurled heads) on the
upper part of the adapter depicted on Fig. 5.2.2.(a) which
is slipped on the drawtube (or the ocular-bearing tube) and
locked into position by tightening the split-tube device.
The adapter may be permanently left in position on the drawtube or on the body of a monocular microscope.

5.2.3.   Focusing with a single-lens reflex camera.

When a camera is fitted to a microscope, the focusing
becomes much more critical than in other types of photography
because of the extremely small depth of focus which is typical of microscope objectives: for example in the green-yellow
portion of the visible spectrum, the depth of focus is of the
order of 8 μm for an objective with a numerical aperture of
0.25 and is less than 1 μm for an objective with a numerical

aperture greater than 0.70. Under these circumstances the
regular ground-glass (ground flat face of a plano-convex lens)
of a reflex camera does not prove very useful: it does not
allow one to focus sharply enough.

Some camera manufacturers provide interchangeable ground
screens; typically they can be obtained with a clear glass,
or a standard ground-glass with a central clear spot (whose
diameter generally varies between 3 and 10 mm). These accessories are excellent for photomicrography, specially those
with a wide clear spot. When they cannot be procured, the
microscopist can transform a standard ground screen in one
of the following ways:

1. a very thin coverslip may be cemented with Canada
    balsam in the central part of the screen (a reticle
    may be conveniently included to mark the center of
    the field);

2. the ground relief can be temporarily obliterated by
    spreading a thin film of a transparent substance.
    Glycerine has often been recommended for this use,
    but it is very hygroscopic and is therefore unsuit-
    able in humid climates. Vaseline is most convenient
    and, as an added benefit, it acts as a dust trap;
    it is easily removed at will by wiping with a piece

of face tissue.

The ground-glass of many 35 mm cameras is larger than the image on the film (nominal dimensions: 24.5 × 36.3 mm) and therefore it may prove convenient to trace with a lead pencil on the glass the contour of the usable field. In the same way the window frame of a 2 × 2" transparency mount may be drawn on the ground-glass; mounts come in different sizes, the most commonly used in laboratories (commercially available cardboard mounts in which a bit of the film is slipped lengthwise) has a 23 × 34 mm window, the longer side being parallel to the length of the film.

5.2.4.   Photomicroscopes.

In the binocular microscope with a photographic attachment, also called photomicroscope or trinocular microscope, a photographic head is mounted in a more or less permanent fashion on the microscope body and the binocular part of the microscope is used as a focusing viewer for the camera.

There are three main variants in the design of a photographic head. The most specialized device, which is also the best one in many respects, is schematically depicted on Fig. 5.2.4.: the photographic chamber and the projecting lens constitute one unit; the lens is large and therefore luminous, and the camera has been stripped down to its bare es-

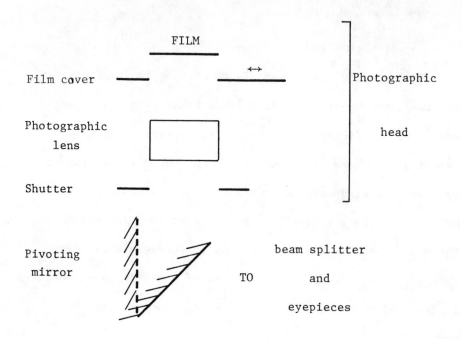

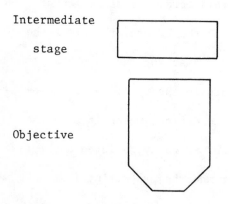

Fig. 5.2.4. Schema of a photomicroscope.

sentials (a film-transport mecanism and a film cover which hermetically closes the chamber and allows one to remove from the microscope the attachment loaded with film). From a mecanical point of view, such a design is very attractive because it is light, which keeps low the center of gravity of the whole apparatus.

Another type of systems makes use of a conventional 35 mm reflex camera as photographic chamber. This kind of set up is versatile, but note that modern cameras usually are cluttered with many gadgets which are of little or no use in microscopical work and only add to the weight; ideally one should use a camera with an aluminium body and as uncomplicated as possible, e.g., a camera body of the post-war vintage.

In a third group of apparatus one uses an eyepiece of standard dimensions as projecting device. This gives more latitude in the control of the final magnification of the image but care should be taken, when changing the projecting eyepiece, that the image in the plane of the emulsion be sharply focused at the same time that the image of the preparation through the observation eyepieces is sharp. This condition is best checked in a darkened room; the back plate of the camera is swung open and a screen is positioned in the plane of the film: for instance of the modified ground-glass viewers described in 5.2.3. or a clear-glass viewer easily constructed by replacing the ground-glass of a 35 mm viewer by a large size coverslip.

As the magnification of the projecting lens is not necessarily equal to that of the observation eyepiece, it is advisable to place in one of the eyepieces a micrometer scale bearing an outline[¶] of the field imaged on the photographic emulsion (see Fig. 7.2.2.).

5.2.5.   Optical configurations.

The spectral characteristics of the light which reaches the emulsion and the desired magnification of the photographic image are the two physical parameters of importance when it comes to select an optical configuration suitable for the photomicrography of a given preparation.

Barring the special case of UV photomicrography, UV components should always be carefully removed from the illuminating beam[§] as their presence would consistently result in blurred pictures; IR rays are best eliminated too. This is

---

[¶]   This outline does not have to be drawn with a very high precision because the magnification of the visually observed image varies with the interpupillary distance.

[§]   Residual UV components will generally be absorbed by the objective (and the intermediate stage) and the eyepiece.

not however the end of the matter and although quite usable images will be obtained with white light under these conditions if one uses apochromatic or good quality semi-apochromatic optical systems, it is often advantageous to use somewhat narrower spectral bands as in particular: i) apochromats are geometrically corrected for two wavelengths only and, ii) losses of light are small with achromats. The use of a limited spectral region is specially advisable when the photomicrographs are to be used for quantitative evaluations[¶].

Projection of an image onto the film by the sole means of an objective is the choice technique when relatively small linear magnifications are sought or when light losses must be kept to an absolute minimum. It follows from an elementary formula of Gaussian optics that the magnification of the image is given by:

[objective magnification]
×[distance (in mm) of the film to the back
                       focal plane of the objective]
÷160   .

In practice, the variation of the projection distance affords an easy means to modify the magnification in spite of the fact that standard objectives are corrected for a projection

---

¶   Conversely a soft image with a pleasant rendition will be obtained using the whole visinle spectrum and an apochromatic objective.

distance equal to 160 mm; but there is a marked vignetting
effect in this configuration and only the central portion of
the object field is usually imaged on the film.

In most instances a projecting element, e.g., an eyepiece, is incorporated in the optical train. In the case of
photomicroscopes with a fixed lens, the investigator is obviously not at liberty to change the magnification (usually
high) of this component, but this is a minor disadvantage in
regard to the high degree of correction which can be brought
to this device and of its intrinsically great luminosity.
When an eyepiece is used to project an image onto the film,
the final magnification of this image can be estimated as
the product of the total magnification of the microscope by
a coefficient equal to the actual projection distance[¶] (in
mm) of the eyepiece divided by the standard projection distance (= 250 mm). But the fact that the magnification of the
photographic image is propotional to the (eyepiece) projection distance is of limited value as 35 mm cameras are not
compatible with very large projection distances as a serious
vignetting effect of the image would then take place, and in

[¶]   This parameter is taken here as the distance between
the back focal plane of the ocular and the plane of the photographic emulsion.

practice the magnification of the image can be varied, on a relatively small range only, by changing the projection distance.

The wide spectrum of eyepiece magnification numbers required is best covered by using oculars of different optical designs:

1. projection eyepieces are low power (e.g., 3X) negative oculars in which the separation distance between the field lens and the eye lens is continuously variable: this feature allows one to vary the distance at which the primary image furnished by the objective is sharply focused. They are very luminous and are most useful for relatively low magnification work;

2. for medium power work, Huyghenian eyepieces are generally preferred because of their relatively high luminosity and fair chromatic corrections; as the eye relief value is of little importance in photomicrography, Huyghenian eyepieces with a rather large magnification number (say 13X, 15X) are quite usable;

3. when eyepieces with a still higher magnification number, say 20 or 20X, are to be used, they should

be of a positive design.

It must be emphasized that certain combinations of objectives and eyepieces will often be found to be well compensated both geometrically and chromatically[¶] : a systematic study of such combinations usually proves rewarding.

A microscope is not ideally suited for very low power photomicrography, that is to say when the final magnification is of the order of 3 to 5X: this is really the domain of macrophotography. However one may have to resort to very low magnifications in two types of situations:

1. no photographic lens is available for the kind of macro work contemplated. It occasionally happens that with very thick specimens the range of the coarse adjustment is too small to permit sharp focusing; in that event the objective may be screwed in the RMS mount fitted at the extremity of the drawtube of a monocular microscope and the focusing is then achieved by moving this drawtube up and down. Note that with this configuration some vignetting of the image is likely to occur;

2. one needs a photographic record of an object at

[¶]   For certain values of the optical length of the microscope and of the projection distance of the eyepiece.

successively higher magnifications.

Needless to say that this kind of photomicrography is carried out using the objective alone (without ocular).

5.2.6.   Positioning the preparation.

Many of the difficulties encountered in photomicrography can be traced back to an inadequate control of the position of the preparation. The great diversity of microscopical objects studied makes it impossible to use one type only of mecanical system to hold them properly; just to quote a few examples, the photomicrography of the pseudopodial movements of an amoeba, of a karyotype preparation and of the interference figure of an isolated crystal will call for different procedures. Two qualities are sought at one and the same time: the preparation should remain steady so that the effect of vibrations of small amplitude for example will be negligible, but it should be easy to adjust. However, from a mecanical point of view, steadiness and easy movability are somewhat conflicting characteristics, specially at high magnification: any microscopist who has tried to "chase" living cells with a standard mecanical stage, using a 40X objective and a 10-15X eyepiece is well aware of the problem!

The most classical positioning device is certainly the mecanical stage which allows one to move the preparation in

two rectangular directions; the smoothness of operation of
this device is much dependent upon its proper lubrication.
Big research microscopes are nowadays equipped with mecanical stages in which the two orthogonal displacements are
controlled by coaxial knurled knobs located near the fine
adjustment control knob. This disposition is very convenient
as it brings close together the controls mainly used during
a manipulation, but mecanical stages built following this
design are not as sturdy as those of the older type in which
the control knobs were close to the plane of the preparation,
and the tightness of their fixation should be periodically
checked.

Although they are sometimes looked upon as too primi-
tive, stage clips may still be advantageously used for low
power work; they can be obtained in a variety of sizes and
two sets at least should be at hand: a pair of short and
rigid clips to secure very firmly a preparation, and a pair
of long and flexible ones which hold fairly well a slide
while allowing one at the same time to displace it with ease.
At times it is expedient to dispense entirely with mecanical
accessories and to secure a slide to the microscope stage
with a small amount of a very viscous medium, for example
vacuum silicone grease. The same technique may be used to

immobilize temporarily Petri dishes. Micro-beakers too can be affixed in this way to a standard glass slide held in place by means of a mecanical stage or stage clips.

A small watch glass set on the circular opening of the stage affords a convenient means to vary the orientation of many specimens, crystals for example; it can be secured to the stage with plasticine (for high power work, metallographic objectives should obviously be used). A procedure of optical crystallography for the study of small well-formed crystals is worth mentioning here as it is susceptible of being applied in many other fields: a crystal is poised (for instance with the help of a very small drop of varnish) on the tip of a glass or steel needle set in a holder which may be nothing more complicated than a bit of cork glued to a preparation slide or a lump of plasticine. If one wishes to examine the sample in a variety of mounting fluids, it is inserted in a microchamber built with cemented coverslips. For low power work, the old "pincette" much in favor with eighteenth century microscopists is still unsurpassed for easily orientating small specimens: it can be constructed from a discarded drawing-pen and can be clamped to the stage or, better, affixed to a ball-joint clamp fixed to the stage.

Tubes are conveniently held in a cradle (see Fig. 5.2.6.).

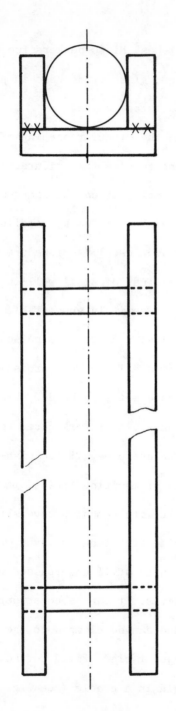

Fig. 5.2.6. A simple brass cradle used to hold test-tubes. On the left, top view; on the right, side view (a test-tube is schematically represented). ×××: solder. In a variant of this design, the two bases of the cradle are not flat as shown here, but V-shaped: thus one cradle can accommodate tubes of different diameters.

Note also that small magnets or magnetic strips may be used to construct extemporaneously various holdings devices.

5.3.   FILMS AND EXPOSURE TIMES.

5.3.1.   Black and white *versus* color.

The selection of one of the two types of emulsions is best accomplished on the exclusive basis of the uses to which the photomicrographs will be put. From this point of view it is expedient to consider two broad classes of utilization: research and demonstration. In critical investigations the overall quality of the image and its aesthetical features are often secondary considerations, and the microscopist must first assess the importance, for the specific study he contemplates, of a number of purely technical parameters: is true color rendition really needed? what degree of reproducibility has to be achieved?... Then he has to face the sad fact that at present the limitations set on the use of color emulsions in microscopical research are numerous:

1. the reproducibility of color emulsions from batch to batch leaves much to be desired, as it has been pointed out many times in the manufacturers' technical literature; they do not keep well, even when stored in a freezer. It follows that for critical work (for instance, involving densitometric evalua-

rions) they have to be frequently calibrated, which
is a tedious and exacting operation;

2. nor do processed emulsions and prints keep well either; some dyes fade with time, and which is worse
   they fade unevenly. This lack of archival permanence
   which can be minimized to a small extent by storing
   the records in a cool dark place, is a great imped-
   iment to long term densitometric studies;

3. high-quality color reproduction in a scientific publication is a very costly process and it is therefore understandable that the editors frown upon its
   unwarranted use;

4. the processing of color emulsions and prints takes
   more time and is more expensive than similar work
   with black and white materials[¶].

The most basic question which confronts the microscopist concerns the exactness with which color rendition is to
be achieved; the answer varies with the nature of the preparation and also with the type of information sought: in the
photography of polarization colors the microscopist will

¶   This is true whether the work is carried out by a commercial outfit or in the laboratory. Note too that rolls of
film sent for commercial processing at times get lost...

clearly strive for true color rendition, but in the photography of a stained histological section he will usually look
above all for good contrast. Thus in the first case the use
of a color emulsion is obligatory, while in the second example one should strongly consider the possibility of using a
panchromatic emulsion and absorption filters.

In a general way then, the use of color emulsions should
best be restricted at the present time, as far as basic research is concerned, to those situations where they prove indispensable and photomicrography in black and white, which
is much more demanding in regard to sharpness of focus[¶],
should be given the preference.

But it is quite a different matter when the photomicrographs are intended for demonstration purposes (e.g., teaching, museum displays). Here what is required is not so much
an exact rendition of the colors but a contrasted image
which attracts the eye and retains the attention. For class
and seminar uses in particular color diapositives have proved
to be most valuable teaching aids[§].

[¶]   So much so that it proves expedient in extreme cases to
photograph with a color emulsion a relatively thick preparation (e.g., of badly formed crystals) and to make black and
white prints. The documents obtained in this way are obviously not of the highest graphic quality, but often very usable.
[§]   But they cannot replace the personal experience of mani-
pulating a microscope. To take an extreme... (contd. p.232)

5.3.2.1.   Black and white films.

There is a large number of black and white films on the market and although each of them has some special characteristics, the microscopist will be well advised to stick to a limited number of them if he does not want to find himself so much engrossed in the photographic aspects of photomicrography that little of his time remains available for microscopical work. Therefore the following list of emulsions has been purposely restricted to a few items which cover a large variety of needs.

All the films listed here are panchromatic. One must regret the lack of orthochromatic emulsions in 135 format (35 mm).

As a very fine grain emulsion, Adox KB14 which has a DIN rating [see Table 5.3.2.1.] of 14° (daylight) can be highly recommended. It is a general purpose film and is specially convenient for crystallographical studies.

Although it is primarily intended for microfilm work,

... (contd. from p.231)   example: how could a student be introduced to the Purkinje phenomenon (*i.e..* the displacement towards the blue of the wavelength of maximum sensitivity of the human eye when the brightness of the field diminishes; in bright light the maximum is around 550 nm and in evanescent light it is *ca.* 507 nm) other than by direct observation?

TABLE 5.3.2.1.

CORRESPONDANCE BETWEEN THE ASA AND DIN RATING SYSTEMS.

ASA	DIN
3	6
6	9
12	12
25	15
50	18
100	21
200	24
400	27
800	30

Kodak High contrast Copy Film[¶] can be useful because of its fine graininess and high resolving power. It has an ASA rating (3200 °K tungsten light) of 64. As is implied by the expression "high contrast", this film is not suitable for recording wide tone ranges. It is very convenient for example for the photomicrography of crystals and for histoautoradiographic procedures, and could prove useful too in darkfield studies. It is available in bulk form and its development is specially rapid: approximately 6 minutes at 20°C with D-19 will permit one to produce crisp negatives. Somewhat softer

¶   Previously called Microfile.

negatives will be obtained by developing with D-76 (1:1).

Kodak Plus-X is a medium speed film (ASA rating: 125, daylight) of relatively low contrast and of moderate graininess. It is a general purpose emulsion, and is available in bulk form. It appears to be used as a "catch-all" medium but one may suspect that in a number of cases the use of emulsions endowed with a lower speed would be practicable: this would result in negatives of better quality.

Kodak Tri-X is the prototype of fast films. Its ASA rating is 400 for daylight, but can be easily pushed to 800 or even 1200 with special developing techniques. It is a low contrast emulsion, even when shot at 400 ASA; because of its graininess it cannot be recommended for routine use in microscopy.

5.3.2.2.   Developers for panchromatic emulsions.

The number of developers available is very large; in addition to ready-made preparations sold by reputable manufacturers and to the formulae they publish, there are many "special" recipes whose claims to fame often border extravagance. Needless to say the microscopist will do well to stay away from the latter and to use only proven methods. It may not be unnecessary to point out that high quality work may be performed without a darkroom (but all the better if

one is available): a changing bag, a daylight developing tank
and the corner of a bench are all that is required.

Two developers formulated by Kodak fulfill most of the
needs of the microscopists. D-19 is a fast-acting developer,
very convenient when one is dealing with a short range of
tones; D-76, and specially its 50% (v:v) dilution with water, is recommended for images with a wide tonal range. They
can be purchased ready-made or prepared in the laboratory
[see Table 5.3.2.2.(a)].

After use, the aliquot of D-19 is returned to its original container. Just before use an aliquot of D-76 developer
is diluted with an equal volume of water, and discarded once
the film has been processed. Both D-19 and D-70 keep up to
six months in well stoppered bottles; if much air is present,
this is reduced to approximately two months. However this
time can be substantially increased if the concentration of
oxygen in the gas phase above the developer is decreased. An
easy way to achieve this result is to hold one's breath for
a few seconds and to exhale slowly in the container, just a-
bove the liquid level, through a pipet. If there is in the
laboratory a supply of some inert gas, the latter can obviously be used with advantage to displace the air within and a-
bove the developer: this is by far the most efficient proce-

TABLE 5.3.2.2.(a).

DEVELOPERS FOR PANCHROMATIC EMULSIONS

Dissolve (all quantities are expressed in g) in approximately 750 mℓ of water at 50°C and in the order given:

	D-19	D-76
Elon	2.0	2.0
Sodium sulfite (dry)	90.0	100.0
Hydroquinone	8.0	5.0
Borax (granular)	–	2.0
Sodium carbonate (monohydrate)	52.5	–
Potassium bromide	5.0	–

After complete dissolution (a hot plate and a magnetic stirrer are helpful but only gentle swirling should be used as hydroquinone is easily oxidized) the volume is made up to 1ℓ (do not use a volumetric flask, a graduated cylinder is sufficiently accurate for this purpose).

TABLE 5.3.2.2.(b).

Approximate length of film as a function of the number of

exposures (35 mm film); trimming and leader lengths have

been included.

Number of exposures	Approximate length of film (in cm)
1	30
2	34
3	38
4	42
5	45
6	49
7	52
8	56
9	60
10	64
15	83
20	102
25	121
30	140
36	163

dure. Soft plastic bottles are becoming popular for storing developers: they are gently squeezed until the liquid reaches the top (thus displacing the air) and then the cap is screwed on. This is an ingenious technique, but care should be taken to check that the plastic used is really inert: in the composition of many soft plastics enter compounds such as plastifiers which can leach out and induce extraneous reactions.

The number of 36-exposures rolls which can be developed per liter of developer is 12 with the D-19 mixture and approximately 7 (using a 280 ml tank) for D-76. The date of manufacture of the various photographic solutions should be noted on their respective containers, as well as the number of 36-exposures rolls processed[¶] (the latter is obviously not necessary if the stock solution of D-76 is exclusively used after dilution, and with stop baths which contain an exhaustion indicator).

5.3.2.3.   Time-temperature development of black and white films.

Black and white films are best developed at temperatures

¶   Or equivalent if, as it is frequently the case, shorter rolls are used. In such instances, it may prove more convenient to note the length of film processed [for an estimation of this length, see Table 5.3.2.2.(b)].

between 15 and 24°C. At temperatures greater than 24°C, standard developers act too rapidly and there exist also serious
risks of damage to the gelatin of the emulsion. Processing at
high temperatures, say between 25 and 40°C, is however feasible if a strong prehardener is used and if sodium sulfate is
added to the developer; but some degradation of the quality
of the negative may be expected to occur in these conditions.
Therefore these procedures should not be used in normal circumstances but best reserved for field work in tropical countries. On the other hand, at temperatures below 15°C, the
washing time becomes prohibitive and archival permanence is
then seriously compromised.

Most manufacturers use the temperature of 20°C as a reference standard. Near this temperature the relation between
the developing time required to obtain a certain contrast and
the temperature is approximately linear. Let there be θ the
temperature (in °C) and t_θ the corresponding developing time
(in minutes). We have

$$t_\theta - t_{20} = \alpha(\theta - 20)$$

where α is a constant (expressed in minutes/°C) characteristic of the developer. Between 15 and 24°C:

$\alpha = -0.6$ min/°C   (= −36 s/°C)   for D-19

and $\alpha = -1.3$ min/°C   (= −78 s/°C)   dor D-76 (1:1)   .

As an example let us consider the following problem: after a few trials it was determined that a satisfactory contrast (for negatives corresponding to a certain type of preparations) was obtained by processing the film at 18.5°C for
11 minutes with a (1:1) dilution of D-76. What would be the
processing time at 23°C?

By substitution in the basic equation it comes:

$$11 - t_{20} = -1.3 \times (18.5 - 20.0)$$

and therefore $t_{20} = 9.05$ min

$$\simeq 9.0 \quad \text{min} \qquad \qquad .$$

Then the developing time at 23°C is given by

$$t_{23} - 9.0 = -1.3 \times (23.0 - 20.0) \qquad \qquad ;$$

thus $t_{23} = 5.1$ min

$$\simeq 5\text{min } 5\text{s} \qquad \qquad .$$

5.3.2.4.   Rapid processing schedule for black and white films.

The following schedule enables one to process in 15-20
minutes (depending on the composition and the temperature of
the developer used) black and white films in small tanks.

Step 1: wash in water for 30 seconds with agitation. This
        moistens the gelatin of the emulsion and therefore
        helps to prevent the formation of air bubbles dur-
        ing the next step; the eventual anti-halo backing

is dissolved.

Step 2: develop with agitation for 5 seconds every 30 seconds.

Step 3: the development is stopped by treatment (with thorough agitation) with a solution (16 mℓ/ℓ) of Kodak Indicator Stop Bath for 30 seconds.

Step 4: fixing and gelatin hardening. With Kodak Fixer and occasional agitation: 2 minutes in the case of a freshly prepared solution but never more than 4 minutes.

Step 5: rinse with water for 30 seconds with agitation. This eliminates most of the hypo.

Step 6: the residual hypo is eliminated by treatment (with occasional agitation) with a solution (1:15, v:v) of Edwall Hypo Eliminator.

Step 7: the products of the reaction(s) which took place during the preceding step are eliminated by a thorough washing in water for 5 minutes.

Step 8: the film is then treated for 30 seconds with a dilution (1:200, v:v) of Kodak Photo-Flo. This results in a drastic decrease of the surface tension of the water layer which covers both sides of the film at the end of step 7.

Step 9:   then the film is dried. In order to carry out rap-
          idly this operation, the film is not vertically hung
          with heavy clips, as is customary; but it is set a-
          long an horizontal light cord, for instance with the
          help of clothes-pins and clips, so that it lies in
          a vertical plane. Within a few minutes all the water
          has dripped away and the film is ready for filing.

5.3.3.   Reciprocity failure.

The exposure required for obtaining a negative of given
density is proportional to the brightness of the image and
to the exposure time, within a certain domain (this is the
reciprocity region); but with exposure times above a certain
threshold, this is no longer true and one then speaks of reciprocity failure: the yield of the photochemical reaction(s)
decreases and a larger exposure time must be used to neutral-
ize the effect of this drop of efficiency.

Most panchromatic emulsions do not exhibit reciprocity
failure phenomena with exposure times smaller than 1 second;
for longer times, correction factors such as those listed in
Table 5.3.3.(a) must be used. It will be noticed that they
are approximate only and given for relatively broad ranges
of uncorrected exposure times; this is all that is needed in
most instances because the latitude of exposure for panchro-

matic emulsions is large: in the reciprocity region, for example, many films can be 100% overexposed and still furnish
usable negatives. To be on the safe side one should, when in
doubt, use three exposure times: the one estimated using the
correction factor of the Table and two others, respectively
under- and over-corrected.

Up to a point a slight overdevelopment can be used to
correct small reciprocity failure effects. But a strong overdevelopment will result in a serious degradation of the image and in particular will increase the graininess of the
negative. In the same way, the use of intensifiers is questionable, as they tend to increase the graininess. Under these
circumstances, the bulk - at least - of reciprocity failure
effects should be corrected at the source, that is to say by
increasing the exposure time, or obviously by using a stronger light source if feasible.

Color films too exhibit reciprocity failure defects, but
the situation is much worse than in the case of panchromatic
emulsions because:

1. most color films for which technical data are avail-
    able show the effect for exposure times greater than
    0.1 second;

2. in general it is extremely difficult to correct the

TABLE 5.3.3.(a).

Approximate correction factors for long exposure times as
a function of the estimated (e.g. with an exposure-meter
or with the help of the formulae given in 5.3.5.) time for
panchromatic emulsions. The corrected exposure time is
obtained as the product of the estimated time by the correction factor.

Estimated exposure time (in seconds)	Correction factor
1 - 2	1.4
2 - 6	2.0
6 - 15	2.8
15 - 35	4.0
35 - 70	5.6

TABLE 5.3.3.(b).

CORRECTION FACTORS FOR THE FILM AGFA CT18.

Estimated exposure time (in seconds)	Correction factor
0.1	1.0
1.0	1.1
10	1.3
100	1.7

effects of reciprocity failure phenomena because the latter are color dependent. In other words in the reciprocity failure region the color balance is different from what it is in the reciprocity region. Although this phenomenon could turn out to be useful for special effects, it certainly restricts the freedom of the microscopist.

There is however one reversal color film which, according to data presented by Pearse, looks most promising: it is Kodachrome-X whose reciprocity region extends up to 1 second. Another color emulsion whose use may be considered for long exposure times is Agfa CT18: it exhibits reciprocity failure effects for exposure times greater than 1/10 s, but they are small. Correction factors calculated from Pearse's data are given in Table 5.3.3.(b).

5.3.4.   Color films.

As it was pointed out above, there is little choice to be had at the present time in color films usable for photomicrography, at least with conventional light sources; because of their large power outputs, flash sources make it possible to use films with narrow reciprocity regions and offer the additional benefit that some color films produce images of lower contrast when the exposure time is very short (of the order of 1/500 s)[¶]: the resulting capability of recording wider tone ranges should prove useful in a number of fields, for example in cytology and in cytochemistry.

As in brightfield microscopy most light sources are used with some filters in order to approximate a crude daylight type of illumination, the selection of color emulsions balanced for 5500°K (daylight type films) seems advisable. amongst the color emulsions available, the microscopist could consider the following reversal films:

1.   Kodachrome-X (ASA rating: 100, daylight) for expo-

¶   But note that with very short exposure times, the actual speed rating may substantially differ from the "standard" rating.

sure times up to 1 second;

2. Agfa CT18 (ASA rating: 50, daylight). With exposure times between 1/10 and 10 seconds, a slight degradation of the image would not be surprising; with exposure times between 10 and 100 seconds, a definite degradation of the image should be expected, but with some luck (backed maybe by the judicious use of light-density colored filters[¶]) this image could still be usable[§];

3. with exposure times no greater than 1/10 second, Kodak High Speed Ektachrome Daylight Type may prove useful because it is fast (ASA rating: 160). Note also that it can be developed in the laboratory with chemicals prepared by the manufacturer (Kodak Ektachrome Film Processing Kit, Process E-4). Transparencies and color prints can be made.

5.3.5. Estimation of the exposure time.

For a given level of illumination and a given optical system, the exposure time is the only parameter controllable by the microscopist, as the aperture is constant. The latter

---

[¶] Sold in photographic supplies houses under the name of "color-balancing system".
[§] For a list of applicable correction factors, see Table 5.3.3.(b).

is always small, and therefore the corresponding F number[¶]
is rather high. On the other hand the value of the F number
as a performance index for a photomicrographic lens (or its
equivalent, for instance a projecting eyepiece) should not
be overemphasized: as the real image projected by the objective (eventually in conjunction with an intermediate stage)
is not extended, the square of the aperture is a better measure of the light gathering capabilities of the photographic
system; in particular, the intrinsic superiority of the wide[§]
photographic head of the better photomicroscopes immediately
follows from this property.

A photometer is a very useful accessory for routine
work. The standard photoelectric exposure meter which is incorporated in the pentaprism view-finder of some single-lens
35 mm reflex cameras is at times sensitive enough to be used
for this purpose, but some preliminary trials will be required to determine the effective F number of the eyepieces
used. Exposure meters specifically designed for photomicrography are commercially available, but the microscopist may
wish to consider the advantages to be derived from the con-

[¶]   This parameter is defined as the ratio of a focal length
to the diameter of the corresponding entrance pupil.
[§]   A typical value of the internal diameter of such an attachment is 41 mm.

struction of one apparatus adapted to his special needs. The
main problem, whether for a commercial unit or a laboratory-
made apparatus, is the selection of a detector with an appropriate spectral sensitivity. Cadmium sulfide photocells,
whose sensitivity spectrum presents a maximum around 554 nm,
have proved reliable and convenient. In many cases too a
selenium barrier cell, which exhibits a large response, will
prove adequate, and specially if it is incorporated in a
current-balance circuit. Both CdS and Se photosensitive elements show effects of temporary fatigue; a few minutes wait
generally is all that is required for detector stabilization.

As in regular photography, the brightness of the image
on a focusing screen is a reliable guide in the estimation
of the exposure time; the same is true of the brightness of
the image in a binocular attachment (provided obviously that
one uses always the same eyepieces). In spite of their apparent primitiveness, these procedures should not be de-
spised; they have served well a few generations of microscopists and anyway, exactly as in regular photography, one must
always interpret with care the indications of a photometer.
For example the results of an integrated light measurement
in the case of a sparsely populated preparation illuminated
with a darkfield system are likely to suggest a too long ex-

posure time; the converse is true in the case of a sparse

population illuminated with a brightfield system.

Once an acceptable negative has been obtained[¶] in re-

producible conditions, it is relatively easy to estimate the

exposure time to use when some optical components are re-

placed by others of different characteristics (e.g., object-

ives switched) or modified (e.g., the numerical aperture of

the main condenser). Chamot and Mason have listed the physi-

cal factors which are to be taken into account in an estima-

tion of the exposure time when a symmetrical brightfield

¶   When, in order to bracket the best exposure time, several exposures are made at constant light intensity, it is
advantageous to select trial exposure times in geometrical
progression for the following reason. The curve which represents the variations of the density D of the negative as a
function of the logarithm of the exposure E (E = It, where I
is the light intensity and t the time) and which is known as
the Hurter-Driffield curve, or H-D curve, has a linear portion. The latter is the region of practical interest. In
that domain:

$$D = D_0 + \Gamma \log_{10} (E/E_0)$$

where D_0, Γ and E_0 are constants. At constant light intensity:

$$D = D_0 + \Gamma \log_{10} (t/t_0) \qquad (t_0 = c^t) \quad .$$

Thus to a repartition in geometrical progression of the exposure times (e.g., systematically doubling the exposure time)
will correspond an arithmetic repartition of the densities,
provided that the range of the exposure times falls within
the linear portion of the H-D curve. Usually with one, or at
the most two, series of trials one can determine a satisfactory exposure time.

trans-illumination system is used, and it is easily deduced

that the exposure time follows the relation:

$$t \propto \frac{(\text{magnification})^2}{(NA)^2_{obj} \times (NA)^2_{cond} \times (\text{ASA rating})} \times$$

$$\frac{1}{(\text{light intensity} \div \text{filter factor})}$$.

In spite of its deterring aspect this formula is easy

to use when, as it is usually the case in practice, only some

of its constructive factors are varied at a time. Let us con-

sider for instance the following example: with a given light

intensity one wishes to photograph a certain type of prepa-

ration using a given projection eyepiece and a given emul-

sion; the distance of projection is constant and so is the

optical length of the microscope. Previous experiments have

shown that for this type of preparation it is advisable to

limit the numerical aperture of the condenser to approxima-

tely 8/10 of that of the objective. With a 10X objective of

numerical aperture equal to 0.45, good negatives were obtain-

ed using an exposure time of 1/10 s. One desires now to use

a 40X objective with a numerical aperture of 0.75. What is

the exposure time required?

To solve this problem, we lump into a proportionality constant (α) all the unmodified factors in the preceding formula. It comes, assuming for a while (but see below) that the filter factor is unchanged:

$$t = \alpha G^2_{obj}/(NA)^4_{obj}$$

and α is determined by the relation

$$0.1 = 10^2\alpha/(0.45)^4 \qquad .$$

Therefore the exposure time sought is given by:

$$t/0.1 = 16 \times (0.45/0.75)^4 \qquad ,$$

and finally it comes $t \simeq 0.2$ s. Note that this is an estimated time which eventually (depending on the nature of the emulsion) will have to be corrected for reciprocity failure effects.

Let us now assume that the first trials, *i.e.*, with the 10X objective, were conducted with the emulsion Adox KB14. What would be the exposure time for a picture taken with the 40X objective on a Plus-X film?

For illustration purposes we shall treat the problem *ab initio*. Again, we lump in a proportionality constant (β) all the terms which appear in the basic formula and which are not changed in the experimental set up. Then this formula takes the form:

$$t = \beta G^2_{obj}/[(NA)^4_{obj} \times (ASA\ rating)] \qquad .$$

The film Adox KB14 is rated at 14°DIN, which is equivalent to 20 ASA as it can be estimated from Table 5.3.2.1. Therefore β is determined from:

$$0.1 = \beta 10^2/(0.54^4 \times 20) \quad .$$

The emulsion Plus-X is rated at 125 ASA. Under these conditions, the exposure time sought is given by:

$$t/0.1 = 16 \times (0.45/0.75)^4 \times (20/125) \quad ,$$

and one obtains $t \simeq 0.03$ s. Obviously one could have obtained the answer more rapidly, as for a given objective (and a given set up) the formula which relates the exposure time and the speed rating of the emulsion takes the simple form:

$$t = \gamma/(\text{ASA rating})$$

where γ is a constant. In the case at hand, γ is determined by $0.2 = \gamma/20$ and therefore the exposure time with the 40X objective and a Plus-X emulsion is given by $t/0.2 = 20/125$.

In the case of the phase contrast method of illumination, an extension of the basic formula may be surmised to have the form:

$$t \propto \frac{(\text{magnification})^2}{(\text{NA})^2_{obj} \times (r^2_{out} - r^2_{in}) \times (\text{ASA rating})} \times$$

$$\frac{1}{(\text{light intensity} \div \text{filter factor})}$$

where r_{out} and r_{in} are respectively the outside and inside radii of the diaphragm annulus.

In the preceding numerical computations, the filter factor has not been taken into account for purposes of simplification; we shall now discuss the influence of this parameter in the two situations of greater interest to the microscopist, namely when switching objectives and when changing the color of the illuminating beam.

The filter factor of an achromatic (neutral) filter whose coefficient of transmission is x% is defined as 100/x; in other words, it is the inverse of a transmittivity. Let us then consider an objective focused on a trans-illuminated preparation: an image of a certain brightness will be obtained. We now switch into the beam another objective with the same numerical aperture (so that the amount of light collected by the two objectives is the same) but which contains an extra optical component, say a cemented doublet. There will be a certain decrease of the image brightness due to losses of light by absorption in the additional element and by reflections at the air-glass interfaces. As a first approximation, we neglect the former and admit, for purposes of simplification, that the various refractions take place at normal incidence: this is certainly a crude approximation in most cases, although it should be observed that in the first

stage(s) of an objective the divergence of the rays is great-

ly reduced. Now, a well known formula due to Fresnel gives

for the loss of light at an air-glass interface under normal

incidence, the expression:

$$R = [(n - 1)/(n + 1)]^2 \qquad .$$

With n = 1.5, the loss is close to 4%; it reaches 5%

for n = 1.6. Many optical glasses have a refractive index of

the order of 1.5, but in order to compensate for the absorp-

tion losses and the effect of the normal incidence hypothe-

sis, we shall adopt the value 95% for the coefficient of

transmission of a single interface; therefore for two inter-

faces the coefficient of transmission becomes 90% and if

there are m additional optical elements, the coefficient of

transmission is estimated (taking into account the fact that

a cemented element possesses only two air-glass interfaces)

in % as $100 \times (95/100)^{2m}$, to which corresponds the filter fac-

tor 1.053^{2m}. Some representative values of this parameter

are given in Table 5.3.5.(a). Note that this estimation of

the filter factor associated with a change of objectives also

applies to phase contrast objectives of type A as the overall

absorption of the phase plate is negligible; on the other

hand it would be necessary to introduce formally the absorp-

TABLE 5.3.5.(a).

Estimate of the filter factor associated with the number m

of additional optical elements in an objective.

m	Filter factor
1	1.1
2	1.2
3	1.4
4	1.5
5	1.7

---

tion coefficient of type B objective phase plates as it is

always large.

Strictly speaking, Fresnel's formula refers to a quasi-monochromatic light; as n is wavelength dependent, R is

wavelength dependent too and the filter is not completely

neutral. But because this effect is weak, the consideration

of an average index of refraction is feasible and therefore

one can treat the filter as achromatic.

In this context, the definition of a filter factor for

a narrow spectral region does not present any difficulty; it

will be noted that this involves only phenomena whick take

place within the filter, in accordance with the customary practice of physics. In photography however the situation is slightly different and the efficiency of the photosensitive emulsion is (more or less explicitly) introduced; in other words the emphasis is shifted from what happens in the filter or at the level of the preparation[¶] to the end-product, namely a picture. Thus photographic filter factors are rather composite parameters but their introduction greatly simplifies the practical evaluation of an adequate exposure time: such a factor may be interpreted as the ratio of the exposure time when one uses a light of a certain color, to the exposure time required to obtain an image of the same density when using a white light, that is to say for the microscopist a light with a daylight-type distribution. As the spectral sensitivity curves of most panchromatic emulsions are pretty similar (except in the tails of the visible spectrum), a general list of approximate filter factors can be compiled [see Table 5.3.5.(b)]. Obviously, no such statement can be made concerning color emulsions and the microscopist must resort to systematic trials in each individual case.

¶   Which most often acts as a filter.

TABLE 5.3.5.(b).

Approximate filter factors to be used, with a light beam filtered for UV rays and a panchromatic emulsion, when passing
from a daylight-type of illumination to a colored light. Note
that: i) the data for blue light are unreliable as there are
marked differences berween various emulsions, and ii) if the
field appears medium or dark red, it is much preferable to
use an emulsion sensitized to IR rays.

Color of the field		Filter factor
Blue	light	1.9
	medium	2.4
	deep	3.6
Yellow	light	1.5
	medium	2.0
	deep	3.0
Green	light	3.0
	medium	4.0
	deep	5.0
Orange	light	4.0
	medium	4.5
	deep	5.0
Red	light	7.0

5.4.   MISCELLANEA.

5.4.1.   Identification of exposures.

Keeping track of what is photographed, and how, can be a nerve-wrecking task, even with a sophisticated log system, as the identification of the various sequences by means of the classical series of blank exposures proves to be rather inefficient (and wasteful). A better system, although by no means an ideal one, is to substitute the image of a stage micrometer for some at least of the blank exposures; anyhow it is always advisable to include in each sequence of photo-micrographs the image of a stage micrometer or equivalent: in many instances this will facilitate the exploitation of the negative at a later date and will also help to remove any doubt about the magnification used.

Note also that there is enough room left in many camera chambers to accept a film-marking device: a small light guide or a tiny optoelectronic element could be used to reg-ister a mark along the film perforations.

5.4.2.   Photomicrographs for publication.

The microscopist should keep aware that because of the copying step essential in the preparation of a half-tone, some losses of tone details are inevitable. Thus it may be advisable to submit to the journals prints whose rendition

has been purposedly altered. In case of doubts, a few trials
(photographic copying of the original print and subsequent
printing) will help to take a decision; such tests are particularly indicated when the photographic demonstration of
the phenomenon under investigation essentially involves the
comparison of various shades of grey. In extreme circumstances the matter should be discussed beforehand with the
editor. Note that most scientific journals request that
prints submitted for publication be made on glossy paper[¶] as
the rendition of details is then much clearer[§].

There are as yet no accepted standards for labelling
and marking photomicrographs, but it certainly makes things
easier for the reader to see in a convenient part of the
picture a dimensioned scale than to contemplate in the legend
a statement of the type "1250X". It must be emphasized also
that due to the change of units and nomenclature which has
been endorsed by a considerable number of scientific bodies,
the old micron (μ) happens to be no longer an approved us-
age; its official appellation now is a micrometer (μm)[†].

[¶]   To be dried say on a ferrotype plate.
[§]   They are more difficult to photograph than prints made
on matte paper; polarizing filters will help to remove unwanted reflections.
[†]   Note that the abbreviation "um" often encountered in
typescripts is incorrect.

If the optical data are important for the critical e-
valuation of the pictures, they should be either listed in
the customary section "materials and methods" or included in
the caption of at least one of the figures (and references
made to it when pertinent). This is specially important when
there is much empty magnification[¶]: a simple statement of
the numerical aperture would prevent needless frustration on
the part of the reader.

[¶]   For examples: see most any textbook of clinical bacteriology.

CHAPTER 6

SPECIAL TECHNIQUES WITH POLARIZED LIGHT
AND PHASE CONTRAST ILLUMINATION

6.1. POLARIZED LIGHT.

6.1.1. Terminology.

Many authors have deplored the lack of consistency and coherence in the terms used to describe the state of polarization of a beam of light. The difficulty really stems from the fact that the electromagnetic field is best described as a tensor but that, according to the case, the discussion is carried out in terms of its magnetic or electric component. Traditionally in optics much emphasis was placed on the magnetic vector; nowadays the key role is attributed to the electric vector, and for example Shurcliff used the expedient of employing exclusively terms related to the latter (e.g., in defining the direction of vibration of a linearly polarized light) and of rejecting terms such as plane of polarization and plane of vibration. This is perfectly feasible and certainly sound from a physical point of view; but it

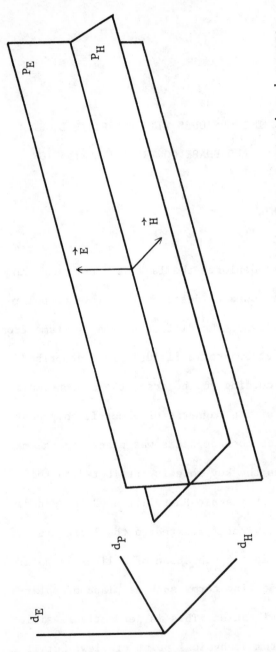

Fig. 6.1.1. Terminology used for the description of light wave propagation phenomena. $\vec{E}$: electric vector; $\vec{H}$: magnetic vector; P_E: plane of vibration of the electric component; P_H: plane of vibration of the magnetic component; d_E: direction of vibration of the magnetic component; d_H: direction of vibration of the electric component; d_P: direction of propagation. P_E and P_H are orthogonal.

departs too much from solidly established customs to be read-
ily accepted by many experimentalists, for example in optical crystallography. Some writers, as Born and Wolf, have
done away with the very ambiguous term of plane of polarization but have kept the concept of plane of vibration; there
is little doubt that the latter is most convenient for de-
scribing the relative orientation of polars and compensators
and therefore will be adopted here. No confusion is likely to
occur if the nature of the component (magnetic or electric)
considered is clearly stated (see Fig. 6.1.1.).

6.1.2.   Polarizing systems.

There are many ways of producing polarized light, but
for microscopical observations only two of them, based on
double refraction and dichroic absorption phenomena respectively, have proved popular. A third glass of polarizing devices, or polars in brief, on which one makes use of reflections at Brewster's angle and which have been employed at
times for microscopical observations, is however worth mentioning as it could still be useful for several special applications.

Polars based on double refraction phenomena and whose
prototype is the Nicol prism (see 6.1.3.) have been satis-

factorily used for more than a hundred years[¶]: they have extremely high extinction ratios and they are achromatic in the
visible part of the spectrum. But they also present a number
of well known disadvantages, the two main ones as far as microscopy is concerned being their rarity in a size large e-
nough to allow one to fill with light the aperture of many a
condenser, and their thickness. The latter is specially troublesome when prism polars are used as analyzers; even if the
lateral displacement of the image which is noticeable in a
Nicol prism is eliminated by using the Glan-Thomson arrangement, the interposition of such polars induces a large variation of optical length which must be compensated when the
analyzer is not in use by the interposition of a thick plate
of optical glass. The necessity of swinging in and out the
correcting glass plate obviously complicates the mecanical
design of the instrument. In spite of these difficulties and
of recent progress accomplished in the construction of linear dichroic filters, prisms polars have still a role in microscopy when strictly achromatic polars are required, for
instance when examining color dispersion of the optic axes

[¶]   Nicol's *princeps* communication appeared in 1829. For a
thorough discussion of the construction of prisms polarizers,
see Johannsen.

in weak interference figures. Another application for which they are nearly irreplaceable is in the study of the (often weak) dichroism of biological preparations.

In certain anisotropic crystals the absorption of light strongly depends on the relative position of the plane of the electric vector and of the crystal axes; under these circumstances the light is preferentially transmitted in one direction. This is the phenomenon of linear[¶] dichroism which was first observed in certain natural crystals and in particular in the tourmalines: a slice of a few millimiters thickness cut parallel to the optic axis[§] of these minerals constitute an excellent polarizing filter as it practically absorbs all the ordinary ray. But such filters are of a very small size and therefore unsuitable for general microscopy. Some organic crystals which can be grown in the laboratory are strongly dichroic too; this is for example the case of the double sulfate of iodine and quinine. This compound was the basis of the herapathite[†] filters which seem to have consisted of oriented crystals embedded in a transparent ma-

---

[¶]   By opposition to circular dichroism: in optically active materials, the absorption is different for right and left circularly polarized light. But the phenomenon is weak and quite negligible here.
[§]   The tourmalines belong to the hexagonal system.
[†]   From the name of the inventor.

trix; but for technical reasons and economy they did not re-
tain the attention of the microscopists. The problem of the
manufacture on an industrial scale of polarizing materials
was solved by Land who showed how to achieve the statistical
orientation of dichroic substances: for example, sheets of
high molecular weight polyvinyl alcohol can be oriented by
mecanical stretching and then treated with iodine. The latter, which is the main dichroic material in use today[¶], becomes oriented through its fixation on the long chains of
the polymer. This type of linear dichroic filter presents
many advantages: in particular filters of a very large size
can be easily produced, and they are thin. The latter characteristic is most important as the interposition of such
devices in an optical instrument does not induce an appreciable variation of the optical length. However these filters have some drawbacks and their lack of achromatism may
prove a serious limiting factor in microscopy: because of
their residual hue, the polarization colors of a preparation
are not shown quite correctly. Therefore for critical work

[¶]   For a review of Land's work on linear dichroic filters
and a discussion of Polaroid filters, see Shurcliff. Note
that "Polaroid" is a trade-name whose unconsiderate use
should be avoided in the scientific literature; for the physicist, a Polaroid filter is a (linear) dichroic filter.

involving the detection of weak color changes, polars of this type cannot compete with prisms polarizers.

Finally, linear polars can be constructed from piles of glass plates: if the parallel incident beam makes with the normal to the plane of a plate an angle equal to the Brewster angle for the particular glass used, it can be shown that the transmitted beam contains an appreciable percentage of linearly polarized light. More specifically let there be α the ratio of the squares of the magnitude of the two components of the transmitted electric vector[¶], which are respectively perpendicular and parallel to the plane of incidence of the light. A somewhat crude formula, in the derivation of which no account is taken of multiple refractions and reflections but which is quite sufficient for the purpose at hand, reads for one plate:

$$\alpha = [2n/(1 + n^2)]^4$$

where n is the refractive index of the glass. As $\alpha < 1$ the perpendicular component is always smaller than the parallel component, and therefore the transmitted light is indeed polarized. With the same assumptions it is easily found that

---

[¶]   The consideration of the squares of the components stems from the fact that the intensity of the light is proportional to the square of the electric field.

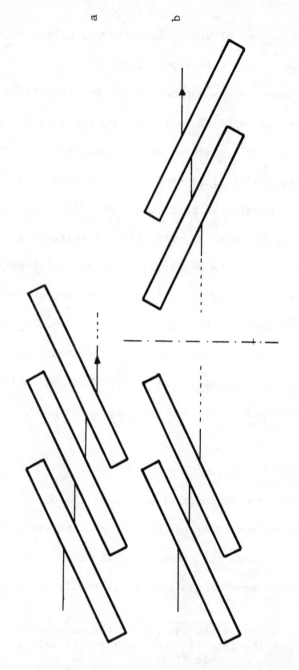

Fig. 6.1.2. Disposition of the plates in a plate polar. In (a), there is an add-itive effect of the individual lateral displacements; in (b), the emergent beam is coaxial with the incident beam.

the ratio of the squares of the two components after passing

through m plates is:

$$[2n/(1 + n^2)]^{4m}$$

With 10 plates, and taking n = 1.50, the latter expression

has for value $0.04^{\P}$. Thus, although one could do much better

with a Nicol prism or a regular dichroic filter, the light

is fairly well polarized. Note that such a polar is achro-

matic.

A device of this type can be constructed from prepara-

tion slides (or of large and thick coverslips); in order to

avoid troublesome interference phenomena, it is advisable to

place spacers (e.g., paper strips) between the plates. Such

polars have been recommended for use in conjunction with

micro hot-stages when the latter are run at high tempera-

tures: they have been found to perform satisfactorily. One

of their disadvantages, something they share with Nicol

prisms, is that they induce a lateral displacement of the

beam. This can be remedied by using two groups of plates

disposed as shown on Fig. 6.1.2., case b; however this con-

trivance is mecanically more complicated to build and to

¶   Note the influence of the value of the index of refraction: with 1U plates and n = 1.52, the ratio of the squares
of the two components becomes equal to 0.03

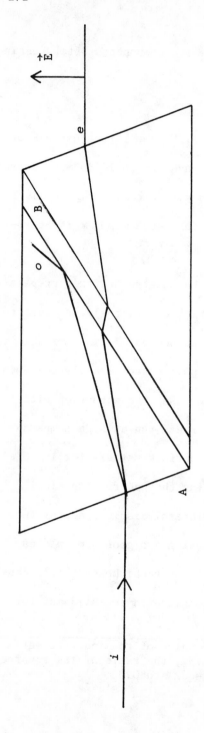

Fig. 6.1.3.(a). Schematic cross-section (not on scale) of a Nicol prism by a plane orthogonal to the rhomb diagonal plane (the latter passes through AB and is perpendicular to the plane of the figure); the space between the two half-prisms is filled with Canada balsam. *i*: incident ray; *o*: ordinary ray; *e*: extraordinary ray; $\vec{E}$: transmitted electric component. The plane of vibration of $\vec{E}$ is identical with the sectioning plane. Note the lateral displacement of the extraordinary ray.

adjust properly than the classical set up depicted on Fig.
6.1.2., case a. It also requires much more space for the same
number of plates, a consideration which is often of importance.

6.1.3.   Determination of the plane of vibration of the e-
lectric vector of a linear dichroic filter.

This determination is most easily carried out with the
help of a polar whose plane of vibration of the electric
vector is known: the two polars are crossed. In a position
of extinction the planes of vibration of the two electric
vectors are orthogonal.

A Nicol prism is ideal for this purpose. It is made of
a natural rhomboid of calcite, a negative uniaxial crystal,
which is cut into two parts along a diagonal plane; the two
sections are cemented with Canada balsam whose index of refraction is intermediate between the two principal indices
of calcite (1.486 and 1.658). In these conditions the ordinary ray is totally reflected while the extraordinary ray
passes through the prism. The position of the electric vector of the exiting beam in relation to the geometry of the
prism is shown on Fig. 6.1.3.(a).

When a Nicol prism or some of its variants are not a-
vailable, one can produce linearly polarized light by reflec-

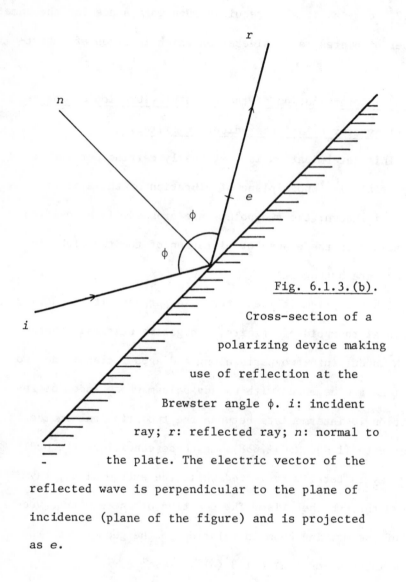

Fig. 6.1.3.(b).

Cross-section of a

polarizing device making

use of reflection at the

Brewster angle ϕ. i: incident

ray; r: reflected ray; n: normal to

the plate. The electric vector of the

reflected wave is perpendicular to the plane of

incidence (plane of the figure) and is projected

as e.

tion at the Brewster angle on a plate of glass, for instance
a preparation slide whose back side has been covered with a
flat black paint: the Brewster angle depends on the refractive index of the glass used (it is equal to Arc tan n) and
is close to 56-57° for common glasses. The main disadvantage
of this device [the position of the electric vector is shown
on Fig. 6.1.3.(b)] resides in its rather low efficiency: the
positions of extinction observed when crossing the polars are
not as sharply defined as with a Nicol prism[¶].

6.1.4.   Test preparations for checking compensators.

Many commercial compensators bear ambiguous markings,
and the wise microscopist will always check carefully the
principal directions of unfamiliar compensating devices. Rec-
tangularly shaped compensators are said to be fast along or
slow along when the slower components (most refracted ray)
is perpendicular or parallel respectively to the long edge
of the apparatus; an arrow is engraved on many compensators
to indicate the "preferential direction", but not all of them
bear an α or a γ stamped on the mount to indicate if it is
the fast or the slow component respectively which is referred

---

[¶]   A reflected highlight is partially polarized too (for
the same reason). This property may be used for orientating
approximately a polar.

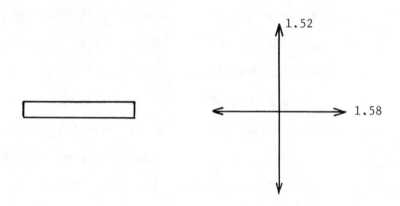

<u>Fig. 6.1.4</u>. Optical orientation of a nylon fiber. The indicated values of the refractive indices are approximate only. Note the use of a customary convention of optical crystallography: the length of a vector is indicative of (theoretically proportional to) the velocity of the light in that particular direction; then the slower component, to which corresponds the larger index of refraction, is represented by the shorter arrow.

to. In the same way many cap compensators bear a small painted mark to indicate the "preferential direction", whose nature must be ascertained. And obviously the principal directions of a laboratory-made compensator must be determined with precision.

For the orthoscopic determination of principal directions, the most convenient auxiliary preparations are likely those of fibers, both natural and synthetic; special mention must be made of nylon fibers, as they are available in a variety of diameters (e.g., monofilament fishing lines) and can be mounted in many different media. The optical characteristics of a nylon fiber[¶] in relation to its geometry are given in Fig. 6.1.4.

For the conoscopic determination of the principal directions of a compensator, one can use as auxiliary preparation a small plate of mica of thickness 0.5-1 mm: mica is a negative niaxial mineral whose preferential cleavage plane is perpendicular to the bisector of the two optic axes. The latter may be simultaneously observed with a dry objective of high numerical aperture.

6.1.5.   Laboratory-made compensators.

Except in specialized laboratories, the construction of compensators is generally limited to flat compensators such as quarter-wave or first order red plates, and to Berek

---

¶   W. McCrone, pers. comm.

compensators if machine shop facilities are available. All these compensators are conveniently made of thin strips of mica, whose (plano)birefringence[¶] is 0.004 for sodium light[§]. Some conflicting reports have appeared in the literature concerning the feasibility of preparing cleavages of mica suitable for this purpose. The origin of the mineral has much to do with its cleavability: physical chemists who need relatively large and well formed strips for the study of film pressures by the Wilhemy plate method have run into this problem and it has been found[†] that mica from Madagascar is more easily cleaved. This is achieved by gently rolling a dissecting needle between the cleavage faces.

Let us note finally that stretched sheets of many synthetic or semi-synthetic materials such as polyethylene and cellophane have been used by a number of investigators as compensators. The main disadvantage of such devices is that their reproducibility is low.

6.1.6.   Microscope configurations.

Microscopical observations with polarized light can of-

[¶]   Also called birefringence on the plate view.
[§]   For purposes of comparison, other typical values are 0.009 for quartz cut parallel to the optic axis and 0.172 for calcite cut parallel to the optic axis.
[†]   J. Guastalla, pers. comm.

ten be carried out in a number of ways, and the classical
problem of deciding upon the acquisition of specialized e-
quipment or upon the adaptation of existing apparatus is best
solved after considering the strength of the phenomena studied and analyzing the types of problems to be investigated.
When the phenomena are weak, the use of a monocular instrument is indicated; on the other hand, the capability of a
given microscope to handle a large variety of works with polarized light is largely function of its basic mecanical
construction. The greatest versatility is obviously attained
with instruments which have been specifically designed with
this purpose in mind: the design of a polarizing microscope
incorporates in particular a rotating stage and slots for
the insertion of compensators and of a Bertrand lens.

The extent to which biological microscopes are usable
for general examination with polarized light mainly depends
on the type of stages with which they are fitted. A few decades ago, many biological (monocular) microscopes were furnished with a rotating and centerable stage as standard e-
quipment and therefore through the addition of a polarizer,
a cap analyzer and a few cap compensators they were easily
used for many studies of chemical microscopy. The modern
biological microscope with its fixed square stage (which al-

lows one to use sophisticated mecanical stages) offers fewer possibilities in this respect, although it can obviously be fitted with a polarizer, a cap analyzer and a cap compensator: it will then permit one to do some limited work. The binocular biological microscope is still less adaptable. With an objective of low power, say no greater than 10X, a linear dichroic filter may be located between the trans-illuminated preparation and the objective. With objectives whose magnification is greater than 10X, the analyzer must be located between the objective and the beam splitter; this may prove difficult to realize in practice, but note that as a makeshift it is always possible to cut a small disk of dichroic material and to position it on the objective diaphragm. The interposition of an analyzer is easier in the case of those microscopes whose revolving nosepiece is mounted on a block adapted to the microscope body by means of a dovetailed slide: it is always possible to insert there a thin sheet of dichroic material.

6.2.   PHASE CONTRAST MICROSCOPY.

6.2.1.   Physical bases and terminology.

In theoretical microscopy one considers two kinds of idealized objects: amplitude objects which absorb part ot the light but do not modify in any way its phase, and phase objects which totally transmit the light but induce some phase

changes. The eye, as well as the usual photometric detectors

such as photographic emulsions and photocells, are sensitive

to changes of amplitude but do not respond to changes of

phase. It follows that the detection of a phase object will

be possible only if variations of phase can be translated

into variations of amplitude. On the other hand, a perfect

microscope objective is one such that the amplitude varia-

tions in the image plane are similar to the amplitude varia-

tions in the object plane; as an image can be decomposed into

a sum of spectra, the transformation of the image of a phase

object into an amplitude image will require the modification

of at least one of the spectral components of the image. This

can be achieved in a number of ways, but with widely varying

degrees of efficiency:

1. as all "real" objectives have only a limited aper-

   ture, some spectra are always excluded from partici-

   pating in the formation of the image and this often

   allows one to detect phase objects, for instance

   single live cells in a biological medium. This ex-

   plains the fact that by closing the condenser dia-

   phragm, that is to say by decreasing in turn the ef-

   fective aperture of the objective, the visibility

of a phase (or of a quasi-hase) object is often en-
hanced;

2. the zero-order spectrum can be eliminated: this is
    the classical darkfield illumination procedure;

3. all the spectra on one side of the zero-order spec-
    trum can be excluded: this is the Schlieren tech-
    nique which as a matter of fact has been little used
    (unjustly?) in microscopy;

4. some other specific modification of one of the spec-
    tral components as it is done in the famous Zernike
    phase contrast method with its two variants of dark
    and bright contrast.

There is no little confusion in the literature (for example in the biological literature) about the meaning to be
attributed to the terms bright and dark contrasts[¶], and as a
result incorrect conclusions have sometimes been drawn as
has been emphasized by Barer. It is fairly easy to see how
one can be led to a faulty interpretation of phase contrast
images. Let us take as ideal phase object a linear grating
of period h, trans-illuminated with a quasi-monochromatic
light of wavelength λ. Its transmission function is of the

¶   Sometimes referred to also as negative contrast and
positive contrast respectively.

form

$$F(x) = \exp[i\phi(x)]$$

where x is a spatial coordinate in a plane parallel to the grating plane (for instance the back focal plane of the objective assumed perfect) and in a direction perpendicular to the grating grooves. The function $\phi(x)$ is real and of periodicity h. We now assume that the phase grating is a weak phase object, that is to say we assume that the phase disturbance induced when it is interposed across the illuminating beam is small. This is to say that

$$|\phi(x)| < 1 \qquad\qquad ;$$

in these conditions

$$F(x) \sim 1 + i\phi(x) \qquad\qquad (1)$$

and the coefficients c_j of the expansion of F(x) into a Fourier[¶] series are:

$$c_o = 1, \quad c_{-j} = -c_j^* \qquad (j \neq 0) \quad .$$

In the Zernike method, one introduces in the back focal plane of the objective[§], a thin transparent plate (phase plate) which retards or advances by $\lambda/4$ the zero order (cen-

---

¶   By definition:

$$F(x) = \sum_{j=-\infty}^{\infty} c_j \exp(2\pi i j x/h) \quad .$$

§   This particular location is selected for reasons of convenience only.

tral order) diffraction spectrum. Then the image is now characterized by coefficients c_j' such that:

$$c_o' = \pm i, \quad c_j' = c_j \quad (j \neq 0) \quad ;$$

this is the set of Fourier coefficients corresponding to the
fictitious amplitude grating

$$G(x) = \pm i + i\phi(x) \quad ,$$

whose light intensity distribution in the back focal plane
of the objective is proportional to

$$|G(x)|^2 \simeq 1 \pm 2\phi(x) \quad . \quad (2)$$

It follows from formula (2) that if the zero order component
is retarded by $\lambda/4$ the phase object will appear brighter
than its background and that if the zero order component is
advanced by $\lambda/4$ the phase object will appear darker than its
background; in both cases the luminosity of the image is a
linear function of the phase change of the object. Under
these circumstances the terms of bright and dark contrasts
are unambiguous.

However formula (2) holds true only in so far as the
approximation (1) is valid; if the grating is a strong phase
object the contrast may be zero or even reversed. In other
words, a dark-contrast image can be obtained with what in
the terminology of the physicist is a bright-contrast set up.
The same conclusion is reached if in place of a linear grat-

ing one considers a non-periodic planar object: the function

$\phi(x)$ is then replaced by a function $\phi'(x,y)$ and the light

distribution of the image will be proportional to $1 \pm 2\phi'$

only if the relation

$$|\phi'(x,y)| < 1$$

is satisfied.

Some aspects of the phase contrast method are more eas-

ily discussed using a slightly different terminology: the

zero order component is spoken of as the undeviated beam

while all the other components (diffraction spectra of any

order $j \neq 0$) are designated by the expression "deviated beam".

Let us now consider more specially the case of a bright-

contrast system. The undeviated beam must be retarded by $\lambda/4$;

this is realized by using as phase plate a glass plate cov-

ered, in the region through which passes this beam, with a

dielectric layer of thickness $0.25\lambda/(n - 1)$. The latter re-

gion which is the optical conjugate of the condenser aperture

diaphragm, is called the conjugate area; the rest of the

phase plate is referred to as the complementary area or the

complementary region. Thus in a bright-contrast instrument

the conjugate area is covered with a dielectric layer; to

advance the undeviated beam by $\lambda/4$ with respect to the devi-

ated beam is equivalent to retarding by $\lambda/4$ the deviated beam with respect to the undeviated beam: in a dark-contrast instrument it is then the complementary area which is covered with a film of dielectric material[¶].

A qualitative analysis of the relationship between the conjugate and the complementary regions allows one to determine the best geometry of the phase plate and of the associated condenser diaphragm. As an objective collects but a fraction of the deviated beam, the quality of the image will depend in particular on the number of the highest (and of the lowest) order of the diffracted component which reaches the phase plate: thus the complementary region should be as extended as possible and the conjugate region should be kept small. This makes it imperative, in turn, that the conjugate area be optically very well defined: this is to say that the construction of the condenser diaphragm, which is imaged on the conjugate region, becomes critical. Now, diffraction phenomena are associated with the presence of a diaphragm and they set a limit to the attainable resolution: as the resolving power is greater with an annular diaphragm than with a circular one, the former pattern is commonly adopted

---

¶   The use of a dielectric layer is characteristic of modern production methods. Early phase plates were made by an etching process.

nowadays. This increase in the power of resolution is achieved at the expense of a loss of brightness of the image, but
in most cases this is of little importance with the light
sources available today.

A most important characteristic of an image is its contrast: in the case of a monochrome image, it can be ascertained from a comparison of the maximum and minimum intensi-
ties. Thus, when a weak linear grating is observed with a
phase plate which retards by $\lambda/4$ the undeviated beam, the
contrast is [see formula (2)] equal to $2\phi/1 = 2\phi$. This value can be greatly increased by absorbing a certain fraction
$(1 - \alpha)$ of the undeviated beam: the phase plate is then
characterized by the transmission function $\alpha\exp(i\pi/2)$ $[\alpha<1]$
and the light intensity distribution of the image is propor-
tional to $\alpha^2 + 2\phi/\alpha$. The contrast has become equal to $2\phi/\alpha >$
2ϕ. As the contrast can be either positive or negative, and
as the absorption process can take place either in the conjugate or in the complementary area, there are four possible
types of phase plate (described in Table 6.2.1. and in Fig.
6.2.1.).

However this enhancement of the contrast substantially
shortens the phase variation range within which an object
can be observed without reversal of the contrast. This is

TABLE 6.2.1.

Preferred notations for the phase-plates used in Zernike phase-contrast method of illumination.

Ab: absorbing layer; Ph: $\lambda/4$ phase changing layer. Note that the symmetry of the table is broken

by the requirement that the conjugate area be very small. See also Fig. 6.2.1.

Conjugate area	Complementary area	Contrast with a weak phase object	Preferred notations
Ab + Ph	nothing	bright	A+
Ab	Ph	dark	A-
Ph	Ab	bright	B+
nothing	Ab + Ph	dark	B-

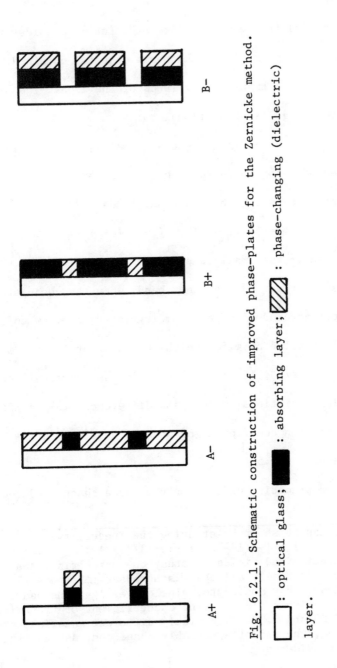

Fig. 6.2.1. Schematic construction of improved phase-plates for the Zernicke method.

☐ : optical glass; ▮ : absorbing layer; ▨ : phase-changing (dielectric) layer.

easily verified in one particular case. Let us consider the
image of (not necessarily weak) phase object Φ obtained with
an A+ phase plate transmitting $100\alpha\%$ of the light. The brightness of the image is proportional to:

$$\alpha^2 + 2[1 - \alpha - \cos\Phi + \alpha\sin\Phi] \qquad .$$

The values of Φ for which the contrast is zero are those
which cancel the square bracket. One finds:

$$\sin\Phi = [-\alpha(1 - \alpha) \pm \sqrt{(2\alpha)}]/(\alpha^2 + 1) \qquad .$$

Let there be Φ_1 the smallest positive value of Φ which satis-
fies this relation. On the other hand, when the conjugate
area is not covered with an absorbing layer, i.e., when $\alpha=1$,
the smallest value of Φ for which the contrast is zero is
$\pi/2$. But $\Phi_1 < \pi/2$[¶]. It follows that contrast reversal occurs
for smaller variations of the path difference; in practice
the transmission coefficient of phase plates is confined in
the range $0.20 - 0.50$[§].

In the preceding discussion the phase change induced by

¶   This can be shown by studying the inequality:
$$1 > [-\alpha(1 - \alpha) \pm \sqrt{(2\alpha)}]/(\alpha^2 + 1) \qquad .$$
§   In specialized fields of study one often finds that one
phase plate is really all that is necessary; for example, Barer concluded that most cytological investigations can be
carried out with a plate having a coefficient of transmission
between 0.20 and 0.35. When widely different materials are to
be observed, the use of a variable phase contrast apparatus
could be considered.

the plate was restricted to the values $\pm\pi/2$; they are specially important because it is under these conditions that
the method reaches its highest sensitivity in the detection
of weak phase objects, but they are not necessarily optimum
in all cases! In extreme situations one must then resort to
the use of a specialized instrument, *i.e.*, a variable phase
contrast microscope (whose discussion falls outside the
scope of this book). But it must be emphasized that it is
feasible to manipulate to some extent the phase changes induced by a fixed plate. Let us consider for example an A
plate with coefficient of transmission α and which retards
by β radians the undeviated beam, *i.e.*, with a transmission
function $\alpha\exp(\beta i)$: the brightness of the image of a (not necessarily weak) phase object Φ is proportional to:

$$\alpha^2 + 2[1 - \alpha\cos\beta - \cos\Phi + \alpha\cos(\beta - \Phi)] \quad .$$

The contrast is a function of β, but the latter quantity is
a function of the wavelength of the light used[¶]. Therefore a
change of wavelength may induce a definite variation of β and
in turn of the contrast. This is most important in practical
microscopy; traditionally phase contrast observations are
carried out with green light as the eye is very sensitive in

[¶]   Directly, and indirectly as the refractive index of the
dielectric layer varies with the wavelength.

this region of the spectrum, but it proves advantageous to examine the preparation with a polychromatic light too, for example with white light: more details will often be revealed.

6.2.2.   Adjustment of the illumination system.

This operation whose importance is critical, is fairly easy to carry out with a well constructed microscope, and in many cases only the initial adjustment will prove to be time consuming.

Step 1. Center the condenser, position a preparation[¶], set up conditions of brightfield illumination and focus the microscope. Then open completely the condenser iris diaphragm.

Step 2. The annulus diaphragm corresponding to the objective selected must now be imaged on the conjugate area of the phase plate. This is not much of a problem as generally the condenser annulus is located in the first focal plane of the condenser and the phase plate in the back focal plane of the objective; as it is shown on Fig. 6.2.2., these two planes are conjugate whatever may be the separation distance of the condenser and of the objective.

Note however that the latter result is obtained within the framework of Gaussian optics and that it is essentially

---

¶   Preferably a test preparation; see 6.2.3.

assumed that the preparation does not introduce any distortion in the path of the light rays; it is therefore imperative to use well corrected (in particular for spherical aberration) optical elements and to select coverslips of the proper thickness. If the annulus diaphragm and the phase plate are not located in the two focal planes mentioned, there will be one position of the condenser for which the image of the annulus on the phase plate will be of maximum sharpness; this is also the case when, whatever may be the location of the annulus and the phase plate, the preparation induces some moderate distortion of the ray paths. A Bertrand ocular is a great help when focusing the condenser, but in the absence of this accessory one can directly observe with the naked eye (after removal of the eyepiece) the plane of the phase plate: this is the equivalent of the Lassaulx method much used in optical crystallography. It will often prove convenient to perform these adjustments with a green light: thus chromatic aberrations are eliminated and the geometrical aberrations are kept to a minimum, which results in a sharper focus.

Step 3. Then the annulus must be centered. Nowadays most manufacturers have adopted a very convenient mecanical design for the condensers used in phase contrast microscopy: the

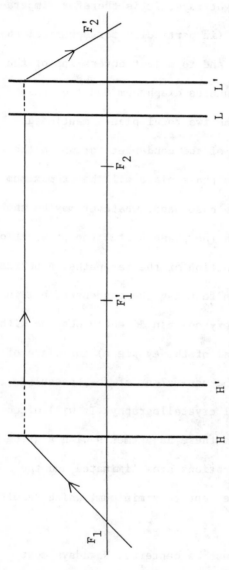

Fig. 6.2.2. On this figure, it is shown that the first focal plane of the condenser (represented by its two principal planes H and H', and its two foci F_1 and F_1')
is conjuguate of the back focal plane of the objective (represented by its two
principal planes L and L', and its two foci F_2 and F_2'): any ray passing through
F_1 will be refracted by the condenser parallely to the optical axis and then
refracted by the objective in such a way that it passes through the second focus
of the latter.

annuli are mounted in a turret so that they can be easily
swung into position. In addition, each annulus is individually centerable[¶] in its fitting by means of two set screws
(working against a spring) which may be actuated by two
spring-loaded screwdrivers.

The instrument is then ready to use. In a well constructed microscope the annuli will stay centered for a long
time, but it is unwise to take it for granted and as a matter
of fact it is always advisable to check, after each manipulation of the turret, that the annulus to be used is properly centered. Many microscopists fail to do so because it is
bothersome to keep switching observation eyepiece and Bertrand ocular. But there is an easy way to obviate the difficulty, which is to employ the Klein method used in optical
crystallography to examine interference figures: keeping in
place the observation eyepiece, one inspects the back focal
plane of the objective through a hand-held converging lens,
for instance an 8X achromatic triplet.

6.2.3.   Test preparations.

They are useful for adjusting a phase contrast illumi-

[¶]   In earlier designs, the whole system condenser-annulus
was laterally displaced with respect to the optical axis of
the microscope.

nation system and indispensable for analyzing the performances of phase contrast objectives. The requirements to be
met by such a preparation are numerous; for an A type objective in particular:

1. the objects should behave as quasi-phase objects,
    that is to say their refractive index should be but
    slightly different from that of the immersion me-
    dium. Although it is by no means mandatory, weak
    phase objects are to be preferred;

2. the coverslip should be of the thickness for which
    the objective has been corrected and should also be
    strictly parallel to the slide;

3. the amount of light diffracted by the preparation
    should be kept to a minimum. It follows that if one
    makes use of a suspension of particles, this sus-
    pension should be very dilute;

4. the shape of the test objects should not be such
    that they appreciably modify the path of the opti-
    cal rays: objects which act as tiny lenses are spe-
    cially unsuitable.

Preparations of small irregular glass fragments mounted
in Canada balsam meet all these requirements. In many instances the following very simple preparation will prove

useful too: a slide is scratched with a diamond point or a
carbide stylus and mounted with Canada balsam.

In special circumstances one will have to relax some of
the conditions listed above in order to test certain objectives under taxing conditions. For example, B type objectives
are expected to be usable with preparations which diffract a
sizeable amount of light: one could therefore test them with
relatively crowded preparations or with periodic objects[¶]
mounted in an immersion medium with a refractive index widely different from their own.

¶   For instance diatoms.

C H A P T E R   7

MICROMETRY

7.1.   PRECISION.

The precision of the measurements performed with a microscope depends on a number of factors, only a few of which are - to any practical extent - under the control of the investigator. For purposes of simplification the discussion will be conducted in terms of linear measurements as they are the ones most frequently carried out; the extension to other types of evaluation (for example, angular measurements) should present no great difficulties.

The limit of resolution ε of the objective used is a lower limit of the absolute error associated with one individual linear measurement. Therefore all the parameters which affect the magnitude of ε, for instance physical parameters such as the numerical aperture of the objective, will play an important role in micrometry. As ε is of the form:

$$\varepsilon \propto \lambda/(NA) \qquad ,$$

one expects at first sight a decrease in the limit of resolution, that is to say an enhancement of the precision of

the measurements when the light used is of shorter wavelength.

This is not necessarily the case however because the value

of the coefficient of proportionality in the preceding for-

mula depends, amongst other things, upon the sensitivity of

the detector used. For the human eye, it is markedly function

of the light intensity[¶] and the situation becomes rather com-

plex. As a general rule, a low level of illumination is to

be preferred as the eye pupil will then be fully open.

But this is not the end of the matter and the limit of

resolution ε will be a reliable estimate of the minimum ab-

solute error associated with one measurement only if the ob-

ject under investigation is either very thin or bounded by

vertical surfaces. In all other cases various phenomena of

refraction and reflection are expected to take place, and in

an unpredictable manner as the shape of the object is usually

unknown. As these effects involve the refractive indices of

the object and of the mounting medium, they are color depen-

dent and they occur whatever may be the type of illumination

method (brightfield, darkfield, phase contrast) used. In

other words, the minimum value of the error on a single iso-

¶   Purkinje's phenomenon.

lated linear measurement is somewhat uncertain.

On the other hand the theoretically attainable relative precision, that is to say the relative precision attainable in the case of a very thin or vertically bounded object, sharply varies with the size of the object. Let us consider the case of a rod-shaped element 10 μm in length observed in green light; if one optimistically assumes that the limit of resolution is of the order of 1 Airy unit, it will be found[¶] that the theoretically attainable relative precision is only 10% if one makes use of an objective with a numerical aperture between 0.30 and 0.35, and that it will reach 2.5% only if one uses an objective with a numerical aperture equal to 1.35.

It then appears that the precision of micrometric measurements is intrinsically low in the case of objects whose length is smaller than 10 μm. In this range, micrometric procedures are best restricted to the comparison of images obtained in identical conditions.

7.2.   MICROMETRIC DEVICES AND ACCESSORIES.

7.2.1.   Stage micrometers.

They are essential for the calibration of a microscope

[¶]   Using the data of Table 3.2.4.

and usually consist of a fine linear graduation engraved on
a glass plate: typically 2 mm divided into 200 equal parts.
An excellent substitute for this kind of device is a high-
quality[¶] hemocytometer, which proves versatile but whose
modern version cannot be used with phase contrast illumination systems because of the presence of a concave base: this
design (which presents the great practical advantage of dras-
tically reducing the probability of scratching the base)
violates one of the fundamental hypotheses made in the calculation of a phase contrast system, *viz.*, that the preparation is limited by two parallel planes perpendicular to
the optical axis of the microscope.

Stage micrometers with special patterns can be made in
the laboratory using the techniques described in 7.2.3.;
they are calibrated by comparison with an accurate stage
micrometer of a standard design.

7.2.2.   Micrometer eyepieces.

Micrometer eyepieces bear a scale (not necessarily a
linear one) in - or very close to - their front focal plane.
It is often stated that it is preferable to use positive

¶   *I.e.*, purchased from a reputable manufacturer who uses
a master ruling whose calibration is traceable to a good me-
trology laboratory.

eyepieces for this purpose as the micrometer scale is then

exterior to the body of the eyepiece: under these circum-

stances small aberrations introduced by the eyepiece optical

system will equally distort the image of the primary image

and the image of the scale. Furthermore it is claimed that

the easy mecanical access to the plane of the scale is spe-

cially convenient for the construction of filar micrometers.

These arguments are debatable as in particular:

1. well corrected negative eyepieces up to (and includ-
    ing) the tenth power are currently available;

2. crossline reticles and micrometer scales made by a
    photographic process are easier to use with a nega-
    tive eyepiece: they are then automatically protected
    from dust and mecanical abrasions;

3. the argument concerning the accessability of the
    front focal plane¶ is valid to a certain extent in
    the case of a filar micrometer used with a monocular
    microscope fitted with an adjustable drawtube, but
    it has little value as far as binocular microscopes
    are concerned as the plane of the primary image

¶   One may suspect that this argument was borrowed from as-
tronomy: it is plainly valid in the case of refracting and
reflecting telescopes.

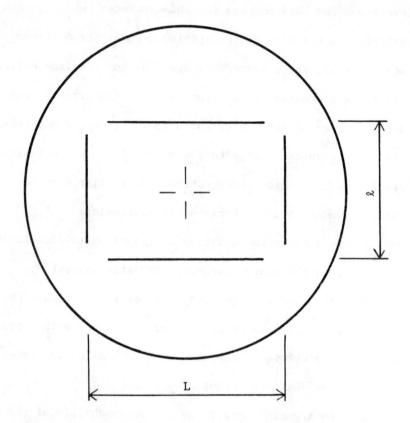

Fig. 7.2.2. A typical field micrometer used with positive eyepieces (e.g. 11X) in photomicroscopes. Let
there be G_P and G_E the magnification numbers of the
photographic lens and of the micrometer eyepiece respectively; then: $36.3/L = 24.5/\ell = G_P/G_E$, where L and
ℓ are expressed in mm.

(for an optical length of say 160 mm) is always well

below the top of the ocular-bearing tubes.

As a matter of fact, many micrometer eyepieces with 10X

magnification have been constructed following a negative de-

sign and they have repeatedly proved to be satisfactory; for

magnifications greater than 10X, positive eyepieces are gen-

erally used for the reasons mentioned in 3.3.7.

Micrometer scales are commercially available in a number

of designs, and the number of special patterns which can be

produced in the laboratory¶ is virtually unlimited. The three

most classical patterns are:

1. a linear scale of 10 mm divides into 100 equal parts;

2. a crossline reticle;

3. a rectangular field outline for photomicrography

   (see Fig. 7.2.2.).

The reading accuracy of a linear fixed scale is severe-

ly limited and this is specially detrimental when it comes

to measure elements whose linear dimensions are smaller than

a few divisions. In such a case filar micrometers prove use-

ful; they come in several types which all derive from a com-

mon design. In the classical version, a movable thin reticle

¶   See 7.2.3. and 7.2.4.

moves perpendicularly to a fixed linearly graduated scale:
the displacement of the reticle is commanded by a micrometric
screw to which is affixed a graduated drum. Things are generally arranged so that one complete rotation of the drum
corresponds to one division of the scale, that is to say the
length of one division is exactly equal to the pitch of the
micrometric screw. However one may dispense with the presence of a micrometer scale within the eyepiece, by adopting
the system used in micrometer calipers. Then the cylindrical
drum is replaced by the beveled edge (graduated) of a thim-
ble whose linear displacement is measured along a scale engraved upon a sleeve. The main advantage of this variant
seems to be that one can make use of parts or sub-assemblies
easily available.

Whatever may be the type of filar micrometer used, it
is most important to perform the measurements in such a way
that the reticle always move in the same direction, say from
left to right, while taking a reading: this is the only way
to eliminate the influence of the inevitable backlash of the
lead screw[¶].

[¶]   This is to say that measurements carried out with filar
micrometers tend to be time consuming. Therefore, say in the
study of a mixture of discrete constituents, a filar micrometer will be used to measure accurately ..(contd. on p.307)

All micrometer eyepieces, either positive or negative, are focusable on the micrometric scale (or the reticle in the case of filar micrometers without internal scale or of crossline reticles). This operation is best carried out by removing the eyepiece from the microscope and looking through it at a distant and extended light source, for instance the sky.

7.2.3.   Fabrication of micrometric scales.

At times the microscopist has to make special scales, for instance to be inserted in the diaphragm plane of an eyepiece or to be projected by the condenser in the plane of the preparation. These accessories are easily made by a photographic process and although they rarely attain the pleasant superficial appearance of scales engraved on glass, they are quite serviceable and the investigator can obviously tailor them to his own needs.

The first step consists in making an accurate drawing of the scale. The degree of enlargement to use depends on a number of factors, the main ones being the optical characteristics of the camera available and the dimensions of the usable portion of the image; for instance in the case of a

---

(contd. from p.306)   a few particles, while a size distribution will be established with the help of a scale micrometer.

disk micrometer supported by an eyepiece diaphragm, the inside diameter of the latter is a limiting factor.

The original drawing should be made on good quality
tracing paper or on tracing cloth, and inked with a full
strength India ink (and usually fine lines). It is then photographed with the help of a conventional copy-stand. The a-
lignment of the camera is critical: the plane of the drawing
(which is advantageously fixed with masking tape to a drawing board covered with a black matte paint) must be strictly
parallel to the plane of the film. As the latter is parallel
to the back of the camera, it is easy to adjust (and check
frequently) its orientation with the help of a spirit level:
the use of this device is indispensable, as small deviations
from parallelism would result in a serious distortion of the
photographic image but go unnoticed on the ground glass of a
reflex viewer. The modified ground glass whose use is ad-
vised¶ for taking photomicrographs is perfect for this type
of work too.

Some thoughts must be given to the selection of an adequate copying film, as the nature of the latter may influence
the style of the drawing. If one uses a conventional high

¶   See 5.2.3.

contrast panchromatic emulsion, such as Kodak High Contrast Copy Film developed with D-19, one will obtain a negative reduction of the drawing. If a positive is required, several ways are opened:

1. one can make a contact print of the negative on a strip of Kodak High Contrast Copy Film (the emulsions should be placed facing each other). If a darkroom is not available, one could rephotograph the negative;

2. one can use a reversal panchromatic film, such as Kodak Panatomic-X, for which a processing kit (Kodak Direct Positive Film Developing Outfit) is available. Note that the ASA rating of this emulsion becomes 64 (for 3200°K, tungsten light) when it is processed for reversal;

3. the drawing can be rendered as the negative of the intended image and photographed on Kodak High Contrast Copy Film.

Once the film has been processed, it should be carefully inspected with a magnifier, for sharpness of contrast. If the image appears satisfactory in this respect, further checks and calibrations can be performed on an image enlarged by projection, or one can proceed directly to the next step and carry out the calibration on the finished product.

Then the film is cut to size; the internal diameter of a standard eyepiece is of the order of 21 mm but it is advisable to measure individually each eyepiece with calipers. There is no need to mount micrometer disks to be used with negative eyepieces; scales intended for use with positive eyepieces or for projection in the plane of the preparation by the main condenser are best mounted between glass plates. The micrometers can then be calibrated under the microscope.

7.2.4.   Construction and repair of crossline reticles.

These operations are relatively easy. Essentially, one affixes to a diaphragm, by means of a suitable cement (a drop of shellac, of household cement, etc...), a thread of very fine material.

The traditional material used for this purpose, and in some respects the best, is spider web. Ancient microscopists were so particular about the characteristics of the threads they employed, that certain species were specified! However most domestic species will contribute material quite useful in an eyepiece[¶]. To collect a spider web, one sweeps it on a piece of light cardboard; all other procedures will result

---

[¶]   Gossamer is plentiful, but very difficult to work with.

in an unmanageable mess. A piece of thread of the desired

length is cut off and positioned with fine forceps and dis-

secting needles.

One should also consider the possibility of using other

materials, more rigid. Quartz fibers are often used in phys-

ics apparatus; their visibility can be increased by plati-

nization, and they are then known as Wollaston fibers; their

use is convenient in the case of a weak image on a dark back-

ground. On the other hand, astronomers seem partial to the

use of reticles made of very fine annealed tungsten wire

(which presents the advantage of being commercially avail-

able). The manipulation of quartz fibers and metallic wires

is made easier if they are kept under moderate tension: this

is customarily achieved by fixing with cement some of the

material to a light wooden bow. It then becomes a simple mat-

ter to position the fiber or the wire, and to affix it to the

diaphragm with two drops of cement. Then the thread is cut

to size.

7.3.   TECHNIQUES.

In a sense there is no microscopical observation which

does not involve, to some extent, a micrometric evaluation as

the investigator automatically relates the size of the object

under scrutiny to the diameter of the field. If the latter

has been previously measured, the main dimensions of an object can be estimated with a precision of the order of 5-10%.

The basic operation of micrometry consists in the determination, for a given optical configuration, of the value of one division of a linear scale incorporated in the eyepiece. This is done by comparing the absolute value of a known length (for example engraved on a stage micrometer) with the length of its image in the first focal plane of the ocular[¶]. In the special case when the length of the eyepiece micrometer scale is known too and the latter is used with a positive ocular, one directly obtains a measure of the magnification of the objective for a certain optical length.

A favorite procedure of microscopists who routinely carried out linear measurements was to adjust the optical length of a monocular microscope fitted with an adjustable drawtube so that each division of the eyepiece micrometric scale corresponds to an integral number of micrometers. This certainly makes for an easier calculation of the results and occasionally this procedure may still prove useful, but it should be kept in mind that a rash modification of the optical length of the apparatus may induce serious aberrations and

---

¶   As it will be easily verified in the case of Huyghens and Ramsden eyepieces (see Figs. 3.3.6. and 3.3.7.).

in particular provoke the appearance of a most troublesome

spherical aberration. In the same vein it must be recalled

here that the optical length of a standard binocular micro-

scope varies with the interpupillary distance but that in

certain instruments this variation is small and therefore

negligible except in critical cases. Each apparatus should

be individually tested in this regard.

In laboratories where micrometric procedures are fre-

quently carried out, it may prove advantageous to dispense

with the use of micrometric eyepieces and to project a suit-

able scale in the plane of the preparation[¶,§]; if needs must,

the projected image can be calibrated in two steps: with a

fixed optical configuration, one first calibrates a microme-

ter eyepiece and then uses it to measure the projected pat-

tern. The projection technique is most versatile when one

uses an external light source, which is best fixed, as well

as the scale, on a small auxiliary optical bench. Microscopes

with a built-in illumination system are not specially well

adapted for the implementation of this technique. Because

they are usually designed around a Köhlerian system, an ideal

---

¶   For a summary of the avatars of this technique, see Chamot and Mason, p.398, footnote 18.
§   See also 1.

position for the scale would be close to the field diaphragm. However this location is unfortunately not very accessible. One could consider the possibility of building a small scale holder to be slipped on the nose of the auxiliary condenser.

Measurements are often made on photomicrographs; in many instances a negative can be exploited directly, that is to say without printing. An enlarger is convenient for this purpose, but before engaging in critical work it should be checked carefully and the quality of the projection ascertained. In the first place, the optical axis of the apparatus must be exactly perpendicular to the image plane (baseboard or screen). Then sharpness of focus and lack of distortion can be checked with a test negative, for example a square network. Note that:

1. even if some distortions are present, the instrument can often be used for certain measurements. By projecting adequate scales or standard images, perhaps simultaneously with the negative to be exploited, matching can be obtained or estimations of the corrections necessary may be made;

2. much work will be eliminated if the negative includes the picture of a square network;

3. when one has to carry out series of linear measure-

ments for which a high precision is not required,

one can take advantage of an idea due to Jelly[¶] and

project not only the negative images but also the

perforations: after an initial calibration, their

geometrical dimensions and their spacing can be used

as references.

Two additional features of large microscopes occasional-

ly find some use in micrometric procedures. The micrometric

calibrated screw which controls the vertical displacement of

either the bodytube or of the stage can be used to measure

the vertical distance (if one knows the refractive index of

the immersion medium) between two points. In the case of

standard preparations, the precision is poor, even if one

makes use of an objective with a small depth of field. Final-

ly, for the estimation of dimensions which exceed a few mil-

limeters, one can use the combination microscope - mecanical

stage as the reverse of a traversing microscope: the object

is fixed to the mecanical stage and is moved across the

field. The displacement is read on the graduated scales (pre-

viously calibrated) of the stage. As these scales can be read

---

¶   Jelley's microspectrograph was disposed in such a way
that the pitch of the film perforations (135 format) corre-
sponded to a dispersion of 50 nm.

only within 1/10 mm, no great accuracy can be obtained and
the system cannot compete with a specially designed apparatus. In all these measurements the influence of the backlash
should be eliminated by always taking the readings along the
same direction.

In addition to linear measurements angular ones are often carried out, as certain angle values are of interest *per
se* while others are needed to compute various geometrical
quantities. Angular determinations can be done, for instance:

1. directly on a photomicrograph or on a projected im-
    age with a protractor;

2. with a goniometer eyepiece;

3. with a graduated rotating stage and a crossline
    reticle. The apex of the angle to be measured is
    brought to the center of the field, for instance
    with the help of an auxiliary mecanical stage mount-
    ed on the rotating stage; then the two sides of the
    angle are brought in succession into coincidence
    with one arm of the reticle and readings taken. The
    angle sought is equal to the difference (eventually
    modulo $k\pi$) between the two readings;

4. with a graduated circular scale located in the front
    focal plane of the eyepiece or projected in the

plane of the preparation by the main condenser.

Group 3 procedures are intrinsically the most accurate, followed (more or less closely, depending on the mecanical construction of the goniometer eyepiece) by those of group 2; group 4 procedures are specially rapid.

In an angular measurement, what is really estimated is the value of the projection of the angle on a plane (H) perpendicular to the optical axis of the microscope. If the plane of the angle and the plane H are not parallel (a fact which may not be apparent if one works with an objective whose depth of focus is too large, and which is rather difficult to recognize on a photomicrograph), serious errors are not unlikely. This illustrates well what is probably the major pitfall of micrometry, namely to accept at face value any experimental result. A critical analysis of the experimental conditions adopted helps much to reduce the probability of recording spurious results. When it is permissible, some manipulation of the sample (for example a change of orientation, a modification of thickness or of degree of dispersion) will often prove useful too.

APPENDIX I.

Standard wavelengths.

Common name	Notation	Wavelength (in Å)
	*	10,140
C line	C	6563
Sodium doublet, sodium yellow, sodium light	D or d	$5893 < \begin{matrix} D_1 \ 5890 \ (s) \\ D_2 \ 5896 \ (m) \end{matrix}$
Mercury green line	e	5461
(Fraunhofer) F line	F	4861
Blue violet mercury line, mercury blue line	E	4358
Mercury line in the U.V. region	**	3650

APPENDIX II.

Construction of diaphragms and stops.

Special diaphragms and stops are easily made in the laboratory:

1. with painted cellophane sheets (see 4.5.6.);

2. of metal;

3. photographically (see 7.2.3.);

4. by a technique which has been much used in the heroic times of phase contrast microscopy to produce diaphragms with annuli, crosses, etc... and which is worth recalling: a transparent disk was coated with fine black particles (the soot of a candle is most convenient) and the desired openings were gently scratched in the black layer. Note that this kind of diaphragm may be given some permanence by fixing the soot with a light varnish in the same way as artists fix a charcoal drawing.

REFERENCES.

Barer, R., "Phase contrast and interference microscopy in cytology", in Physical techniques in biological research, Oster, G. and Pollister, A.W., eds., Academic Press, New York, 1956, 2, 29-90.

Born, M. and Wolf, E., Principles of optics - Electromagnetic theory of propagation, interference and diffraction of light, Pergamon Press, Oxford, 2nd. (rev.) ed., 1964.

Chamot, E.M. and Mason, C.W., Handbook of chemical microscopy, John Wiley & Sons, New York, 2nd. ed., I, 1954.

Gage, S.H., The microscope, Constock Publishing Co., Ithaca, 1936.

Herzberger, M., "Geometrical optics", in Handbook of physics, Condon, E.U. and Odishaw, H., eds., McGraw-Hill Book Co., New York, 2nd. ed., 1967, 6-20 to 6-47.

Johannsen, A., Manual of petrographic methods, McGraw-Hill Book Co., New York, 1918.

321

Paul, J., _Cell and tissue culture_, The Williams and Wilkins Co., Baltimore, 2nd. ed., 1960.

Pearse, A.G.E., _Histochemistry, theoretical and applied_, J. & A. Churchill Ltd., London, 2nd. ed., 1960.

Shurcliff, W.A., _Polarized light_, Harvard University Press, 1960.

SUBJECT INDEX.

(See also the Table of contents.)